Office Ergonomics and Human Factors

Office Ergonomics and Human Factors

Practical Applications

Second Edition

Céline McKeown

CRC Press is an imprint of the
Taylor & Francis Group, an **informa** business

CRC Press
Taylor & Francis Group
6000 Broken Sound Parkway NW, Suite 300
Boca Raton, FL 33487-2742

Printed on acid-free paper

International Standard Book Number-13: 978-1-4987-9910-2 (Hardback)

Visit the Taylor & Francis Web site at
http://www.taylorandfrancis.com

and the CRC Press Web site at
http://www.crcpress.com

This book is dedicated to June and Gerard McKeown, from whom I inherited drive and determination.

Contents

Preface

Ergonomics, or Human Factors, has moved away from being viewed simply as an academic subject, into being recognised as a useful tool in generating safe, comfortable, and productive working environments. This process has been assisted by Ergonomics and Human Factors being presented in a more usable format by authors of Ergonomics and Human Factors texts, as well as Ergonomists/Human Factors Consultants working on the coal face. Because of this change in perception, the responsibility has landed in the laps of departmental managers, facility managers, safety managers, occupational health departments, etc. to pick up the threads of basic Ergonomics and Human Factors principles and apply them in their own working environments.

Computer users face a series of problems once they enter the 'workplace', whether that be an office, conference room, hotel bedroom, plane, or train station. These problems are likely to make them unhappy and uncomfortable, at the very least, and possibly injure them, if no form of control is exerted by those around them. The intention behind this book is to provide practical, usable advice that can be applied directly to any environment where a computer or mobile device user is required to work. The advice has grown from years of first-hand experience of many types of settings and dealing with their inherent problems. It is hoped that the application of the information contained in this publication will ensure that computer or mobile device users no longer have to use unsuitable workstation furniture or equipment; they will not have to perform tasks that are badly designed; and they do not remain ignorant of the very basic advice needed to ensure they can work comfortably and safely.

Author

Céline McKeown is a Consultant Ergonomist with Link Ergonomics, based in Nottingham, England. She works as a consultant to industry, as well as to commercial organisations. She specialises in identifying the causes of injury and ill health in the workplace and designing them out of the system. She also has a special interest in product design. She trains engineers, architects, health and safety professionals, and others who have responsibility for changing the working environment so that they know how to incorporate ergonomics and human factors during the planning stages. She acts as an expert witness for defendants and claimants in work-related personal injury claims. She is a Fellow of the Chartered Institute of Ergonomics and Human Factors in the United Kingdom, and she is a Fellow of the Irish Ergonomics Society.

1

Working Posture

1.1 Introduction

One of the key elements in ensuring that people can work comfortably and effectively is good posture. There is some controversy regarding what constitutes 'good posture'. The traditional view is that an upright posture is the most appropriate. Others suggest that people prefer to lean back in their seats, while some argue that the seat should be capable of sloping forward so that the angle of the hip is increased to reduce the pressure in the lumbar region of the back. Having so much published disagreement among knowledgeable sources does not help those who need to accommodate their workforce. What would be helpful is having an agreed starting point when it comes to posture and applying it with a little flexibility. This chapter aims to provide an understanding of what can assist in creating a comfortable working posture.

1.2 Sitting versus Standing

At one time it was generally recommended that where a task lent itself to individuals sitting down, a suitable chair should be provided that allowed them to do so. There are accepted benefits to sitting over standing: people can take the weight off their legs; greater stability is offered to the upper body; energy expenditure is reduced, as are demands on the circulatory system. It has been the view for some time, e.g. Bridger (2003), that a person who is required to stand for prolonged periods can experience physiological changes, including peripheral pooling of blood, a decrease in stroke volume, and increase in heart rate. However, it has to be understood that this outcome results from extended standing throughout a shift, where the individual performs an activity intended to be completed whilst standing, such as in a factory or an assembly-type environment.

This is not necessarily the case where the individual can alternate between seated and standing desk-based work. Despite the benefits to be enjoyed while sitting, however, there is general agreement that people are not designed to sit for extended periods without interruption. Barbieri et al. (2017) proposed the use of sit–stand desks as part of an initiative to decrease sedentary behaviour among office workers and to reduce the risks of negative effects on their cardiometabolic health. As part of their study, they promoted 10 minutes of standing after every 50 minutes of sitting. The workers reported that the use of the sit–stand desks contributed positively to their health and well-being, without interrupting their work.

Once individuals are sitting, many work in a manner that is likely to increase feelings of discomfort. For instance, some people tend to slump forward while working, which can have a negative impact on digestion and breathing. This is one of the main reasons why all people who work while seated should be given awareness training, which includes an understanding of how the back works, the mechanics by which it can be overloaded, and, more fundamentally, what type of posture they should be adopting and how they can actually adjust their chairs. No assumption should ever be made that a seated user knows how to operate their chair and how to make suitable adjustments so that they can work in a suitable, fully supported posture.

1.3 The Back

The back consists of many parts, and any one of these can be subject to general wear and tear, disease, the aging process, and abuse. Abuse of the back in the workplace is very common, but, generally goes unrecognised. There is too little appreciation of how easy it is to subject the back to unnecessary stresses inadvertently.

Apart from offering support to the trunk and being a main agent in its movement, the back also plays a part in steadying the upper limbs and head. This ensures that they can be moved and repositioned smoothly and that they can bear the stresses encountered as the person works.

Figure 1.1 is an illustration of the spine. The spine consists of 33 individual bones, referred to as vertebrae. It is normally divided into several distinct sections: the cervical vertebrae, comprising seven vertebrae in the neck; the thoracic vertebrae, comprising 12 vertebrae that incorporate the chest and rib area; the lumbar vertebrae, formed from five vertebrae in the lower back; the sacrum, which is formed by five fused vertebrae; and the coccyx, which is made up of the remaining three or four rather simple vertebrae. Because the weight borne by the lumbar vertebrae is greater than that borne by either the cervical or thoracic vertebrae, and because the lumbar region is subject

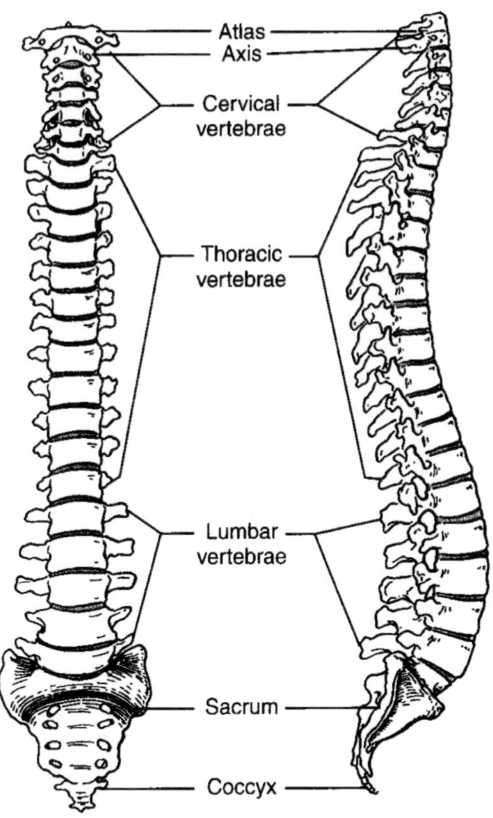

FIGURE 1.1
The spine. (From Watkins 2009.)

to a peak in leverage at this point (as a person leans forward, backwards, or sideways), it is subjected to higher levels of stress.

It has been suggested that 80% of people will experience back pain at least once in their lives. Back pain can occur with or without structural damage. Back pain that tends to be persistent is suspected to be due to structural damage or degeneration (Watkins 2009). Although it has been reported (Zhang et al. 2009) that the source of back pain is often difficult to diagnose, there is a view (Anderson and Tannoury 2005) that disc degeneration was one of the main reasons for chronic low back pain (LBP). Zhang et al. (2009) have a slightly different view and suggest that most data appears to indicate that chronic LBP is most closely related to the anatomical structure of the intervertebral disc, particularly in individuals with no obvious herniation of the disc, and is the disease process known as discogenic lower back pain (DLBP). DLBP is a loss of lower back function with pain.

The discs are like cushions that sit between the vertebrae and act as shock absorbers. They also give flexibility to the spine, allowing the individual

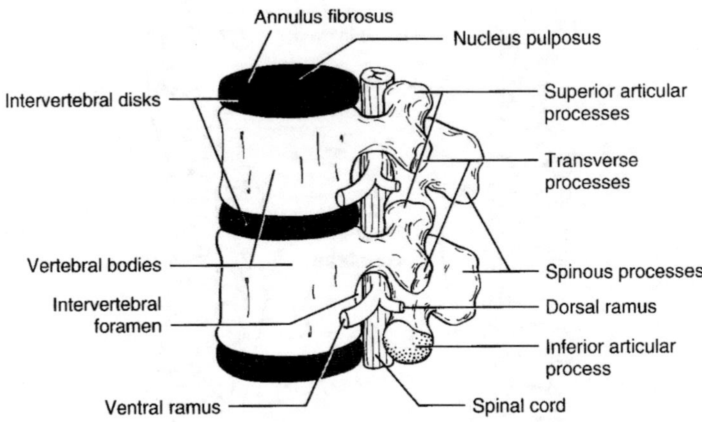

FIGURE 1.2
The discs. (From Watkins 2009.)

to bend sideways (lateral flexion) and lean forward (flexion) or backward (extension). Figure 1.2 is an illustration of the discs.

The discs in the cervical and lumbar regions tend to be thicker at the front than at the back, which contributes to the forward curving of the spine in these areas. The discs also get thicker as they progress from top to bottom along the vertebral column. It is estimated that they make up 20% of the total length of the vertebral column.

The discs are made up of two parts: the annulus fibrosus, a fibrous outer layer, and the nucleus pulposus, a gelatinous mass similar in consistency to toothpaste. Because of its consistency, it can change its shape easily. As a consequence, when a person bends in any direction, the nucleus pulposus becomes wedge-shaped, with the narrow end located toward the direction of the bend. At the same time, the annulus fibrosis bulges on the opposite side of the bend.

Degenerative changes in the discs first start to appear when an individual reaches their twenties (Zhang et al. 2009). When this starts to happen, the annulus fibrosis can bulge or rupture if it is subject to excessive stress, particularly stress imposed by bending. The bulges or ruptures (also known as prolapsed or slipped discs) tend to happen posteriorly because this is the point where the discs are thinner. They also tend to happen more frequently in the lumbar region and the lower cervical region.

A prolapse typically occurs suddenly as the result of an acute failure in the annulus fibrosis. Because the discs do not have their own nerve supply, they are not a direct source of pain. Pain is due to pressure being exerted on a nerve root or spinal nerve by the bulge or rupture and will be experienced in the areas served by the nerves in question. For instance, a protrusion in the lower cervical region may produce discomfort in the hand.

It is not uncommon for an individual who has complained of symptoms in the hands, which are suggestive of an upper limb disorder, to be diagnosed as suffering from a degenerative condition in the neck. Protrusions in the lumbar region can cause shooting pains in the legs, usually starting in the buttock and moving into the back of the thigh, sometimes travelling as far as the feet. This is commonly referred to as sciatica. There is often local muscle spasm following a rupture, which also causes pain.

The muscles of the back—the 'true' muscles—form several layers, which are actually covered by the muscles of the upper limbs. The 'true' muscles are the primary means by which the spine is kept erect and can be rotated. The main task for the muscles when the back is upright is to resist the pull of gravity. If the body is balanced when upright, the back muscles do little work to maintain its stability. If an individual leans forward, the muscles contract to prevent them from falling forward. The muscles also 'manage' the lean so that it is executed smoothly. The farther forward an individual leans, the more active the back muscles become. The same muscles assist in controlling the upward movement as the individual stands or sits upright again. In conjunction with the back muscles, the abdominal muscles are also called on to assist, which explains why people receiving physiotherapy for a back injury often get advice on strengthening the abdominal muscles.

Once an individual adopts a sitting position, they rely on static muscle work to keep them in that position. This requires prolonged and uninterrupted contraction on the part of the muscles. Dynamic muscle work, on the other hand, involves the contraction and relaxation of muscles, which achieves movement. Despite the fact that during static muscle work there may be no discernible movement, which might suggest no work is being done by the body, static muscle work is considered to be more demanding than dynamic muscle work. As a consequence, longer periods of rest are required to recover from this type of work than from dynamic muscle work. Tasks usually involve a combination of static and dynamic work. For instance, when an individual is sitting at a desk performing a computer-based operation, the muscles responsible for controlling the back, shoulders, and arms employ static muscle work while the hands employ dynamic muscle work as they use the keyboard and operate the mouse. Static muscle work can result in minor discomfort quite quickly. People who continue to work without changing position or taking a break can start to experience increasing pain.

The ligaments, in conjunction with the muscles, play a part in stabilising the individual when in an upright position. These can be likened to straps that stretch between bones, and they provide passive resistance and limit movement toward the extreme range of movement of the spine. It is thought that ligaments and muscles are susceptible to injury as a result of twisting and stretching, particularly if repeated over extended periods of time (McKeown 2011).

Given that the posture adopted by the individual has a significant impact on the health of the back, it would seem sensible to ensure that workers are given appropriate information to allow them to work comfortably and reduce the likelihood of them developing injuries.

1.4 Posture

Some studies have purported to demonstrate that, when given the choice, seated individuals choose to sit in anything but an ideal position, preferring instead to lean forward or backward in their seats. However, none of these studies seems to make it clear whether these individuals have received instruction on how their chairs operate and what they should aim to achieve with their posture. As a consequence, they may simply be adopting postures dictated by their degree of knowledge, or lack of it.

Discussing posture is not just about what is right for the back; it also has to take into account of what is right for the upper limbs, the head and neck, and the lower limbs. When it comes to back position, the starting point should be to ensure that it is reasonably upright, which is completely different from suggesting that someone should sit erect. Sitting in a reasonably upright position actually means that the individual may have a slight—very slight—backward lean that positions them at about 110° relative to the base of the seat. This is considered to reduce disc pressure and to reduce the workload for the back muscles because the backrest of the chair offers greater support for the back; an assumption is made at this point that the chair in use will provide the required amount of adjustment to ensure that the individual is supported in the chosen posture rather than having to adjust his or her posture to a position that accommodates the limits of the chair.

Studies have demonstrated that if we did insist that people work in an erect posture, this would likely result in a forward slump. This is due to the fact that the static muscle work required to maintain the upright position is particularly fatiguing, so people relax the muscles responsible for holding them upright and slump forward. This causes deformation of the discs.

Should an individual decide to work in an inclined position that exceeds about 110°, this will require extending the arms farther forward to reach the keyboard or mouse. An important goal of good working posture is to reduce the workload on the arms. The workload can be minimised if the arm position can be kept close to normal or neutral during the course of the work. When the arm is completely relaxed, it falls naturally by the side of the body. The sitting position relative to a desk and its equipment, such as a keyboard and mouse, should ensure that the individual simply raises the forearm so that it forms a 90° angle with the upper arm, and the upper arm should remain in its natural position alongside the ribcage. If the individual sits at a distance from the desk or leans back in the seat, they have to overcome the increase in reaching distances by moving the upper arm forward. Holding it in this position during tasks requires static muscle work, already described as particularly fatiguing. To facilitate the appropriate positioning of the arm, it is essential that the keyboard and mouse be located close to the leading edge of the desk, but not so close that there is no space to rest the hands when not operating the equipment. A 10 cm (4″) gap in front of the keyboard is considered sufficient to rest the forearms in between bouts of keying.

Leaning backward in the chair to an extreme degree will also have a negative impact on head and neck position. This is due to the fact that as the individual reclines, the head tilts back relative to the screen, keyboard, or documents being viewed. This will require a compensatory repositioning of the head. In other words, the head will have to be bent forward, and this is likely to result in neck pain over time.

Task chairs that incorporate a forward tilting seat have been advocated as a means to reduce the need for hip flexion. However, getting people to sit on a slope appears, in practice, to have two likely outcomes: they will brace themselves with their feet to combat the sensation of slipping forward; and women wearing skirts, particularly lined skirts, will find that they tend to slide out of their skirts.

Specifically designed stools with sloping seats have addressed the issue of users slipping forward on the inclined seat by incorporating knee pads that hold the user in place. Studies have shown that users of these types of seats can complain about discomfort in their knees. These chairs also lack any form of back support, and, as a consequence, the back muscles have no opportunity to achieve any type of support and have to maintain the upright posture throughout the time the individual sits on the seat. This requires long periods of static muscle work. This is compounded by the fact that the seat does not allow for easy variability of posture. When these points are combined with the fact that sloping seats are not particularly easy to get into and out of, it might be advisable to permit them to be used only under the instruction and guidance of a health professional such as a doctor or physiotherapist.

The starting point when setting up a seated posture is working height. Assuming a desk of fixed height is in use, the chair should be adjusted so that the individual's elbows are level with the home row (i.e. the middle row) of the keyboard (see Figure 1.3). By sitting at this height, users will not have to raise their shoulders when operating the keyboard or mouse nor move their upper arms away from their sides, as what would occur if they sat too low relative to their equipment. In addition, they will not have to work with their forearms above the horizontal. If the forearms are kept in a horizontal position, with a 90° angle at the elbow, they will be able to work with a straight line running through their wrists and into their hands. This makes them less likely to encounter symptoms of an upper limb disorder (ULD) in this area. (The issue of ULDs will be dealt with further in Chapter 11. ULDs are also known as Repetitive Strain Injury, i.e. RSI.) If a particularly tall individual finds that he or she has to lower the seat to get their elbows level with

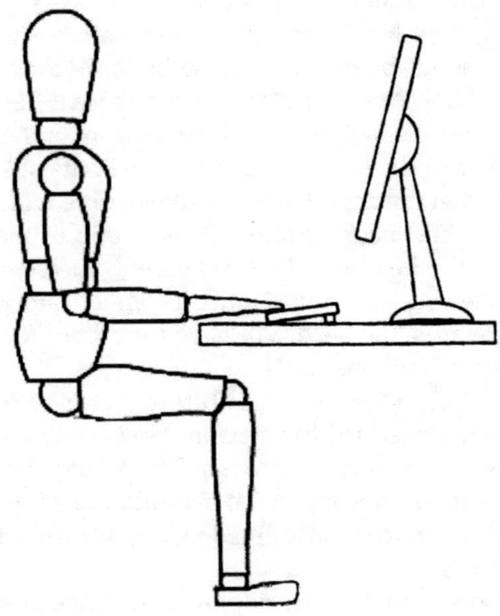

FIGURE 1.3
Ideal starting point for working posture.

the home row of the keyboard to such an extent that their hips are lower than their knees, that person needs a higher desk and probably a higher chair. If an adjustable-height desk is in use, the user would position the chair to get their feet on the floor, then position the desk to achieve the same arrangement as described above.

Once the chair height has been altered, attention should be turned to the feet. If an individual finds that, having moved the chair, their feet are no longer firmly on the floor, they should use a footrest. (The most suitable type of footrest will be discussed in Chapter 2.) The footrest should support the feet at a level where the user is able to work with an approximate 90° angle at the knee (see Figure 1.4). This angle can be increased slightly by sliding the feet (and footrest) forward, but should not be increased by having the chair too high, nor reduced by having the chair too low.

A footrest should not be used by anyone who finds that their feet easily reach the floor once the chair height has been altered. Doing so can have quite an adverse effect on comfort level. This results from two specific consequences. In the first instance, as the foot is raised off the floor and placed on the footrest, the knee is raised upward by a similar amount. As the individual does not change the seated height, the angle at the knee is reduced. This can result in a restriction of the blood flow into the lower legs, and the user will start to feel uncomfortable and fidget without realising why. Second, Figures 1.3 and 1.4 clearly illustrate that a significant proportion of

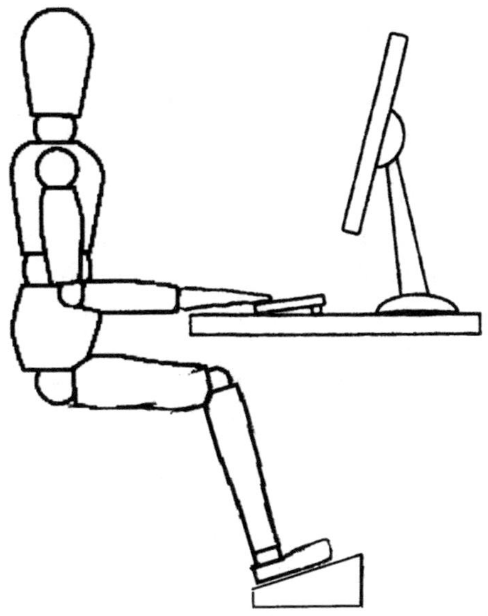

FIGURE 1.4
Using a footrest correctly.

the buttocks and backs of the thighs is likely to be in contact with the seat surface. This will ensure that a large surface area of the body supports the weight of the torso. If the knee is raised, owing to the unnecessary use of a footrest, part of the back of the thigh will also be raised, reducing the surface area supporting the upper body's weight. Focussing the upper body's weight on a smaller supporting surface will result in pressure points, again making the individual uncomfortable and likely to fidget.

If people choose to work without footrests and their feet do not touch the floor, they are likely to encounter other difficulties. The leading edge of the seat is likely to press into the backs of the thighs as the weight of the lower legs drags them down. The compressive effect will create discomfort, ultimately making people fidget.

Once the seated height is established and properly supported, the focus should turn to the back and to the degree of support offered by the backrest of the chair. Initially, individuals should adopt the position they would like to work in for a period of time, whether that be upright or leaning slightly back. The backrest of the chair should be capable of allowing them to adopt such a sitting position and should offer an adjustment mechanism, such as a lever under the seat that can be easily reached from the seated position.

Once the backrest has been adjusted to support the upper body posture of the individual, the lumbar support of the backrest should be positioned so that it falls in line with, and supports, the small of the back. Some chairs may not have this specific facility, and this is discussed further in Chapter 2. It is important that the angle of the backrest is altered prior to positioning the lumbar support, because the lumbar support of a chair will appear to 'move' upward as the individual leans back in the seat, or will 'move' downward as the individual moves to a more upright position; this could leave the lumbar region unsupported.

As Figure 1.3 clearly illustrates, keyboard users should not rest their wrists on the desk or keyboard. Their hands should 'hover' over the keyboard. By doing so, the individual will be able to call upon the larger muscles of the arms to move the limbs and thus reposition the hands relative to a particular key. If people rested the wrists on the desk surface while keying, they would, in effect, be anchored in place and would have to extend their fingers to reach the keys. It is also likely that if people rest their wrists on the desk and raise their hands up toward the keys, they will work with bent wrists. Again, this is known to be associated with the development of upper limb disorders.

Individuals should sit as close to the leading edge of the desk as they wish and should have the keyboard and mouse within easy reach. They should not be prevented from sitting close to the desk by bulky armrests on the chair or by design features of the desk. The aim should be to allow them to work with an approximate 90° angle at the elbow. This will ensure that the upper arm can hang naturally by the side of the body and will demand a minimal amount of work from the muscles. The farther they sit away from the desk, or the farther they position their keyboard and mouse from the leading edge of the desk, the farther they will have to reach with their arms. Working in such a manner will increase the workload for the arms and the rate at which they become fatigued.

Head and neck posture will be determined by the visual requirements of the task. The aim should be to present equipment, such as the screen, at a height where the individual can adopt an upright, but not necessarily erect, head position. Figure 1.5 shows that when an individual sits or stands erect with the head completely upright and looks straight in front of the body,

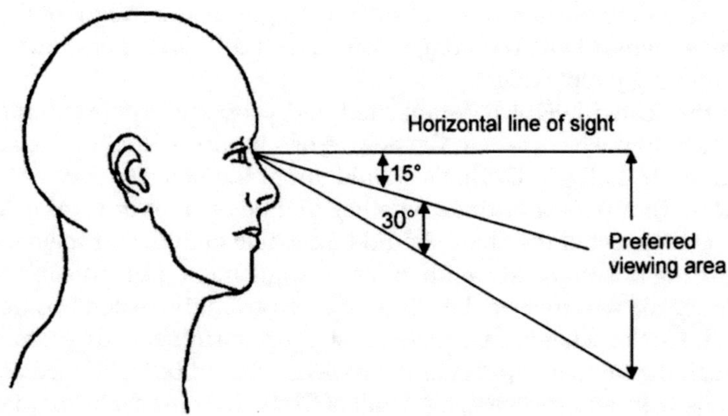

FIGURE 1.5
Illustration depicting the horizontal line of sight and preferred viewing area.

a stance similar to that adopted by a soldier standing to attention, the gaze will follow what is referred to as the horizontal line of sight. However, given the opportunity to relax, the head will drop down slightly because of its weight, and the eyes will ultimately adopt a downward gaze somewhere between 15° and 30° below the horizontal line of sight. This is considered to be the preferred viewing area. People can, of course, view information presented in an area that falls below this preferred viewing area, but this does require the head to be tilted downward to a greater extent. Pheasant and Haslegrave (2006) have pointed out that the neck muscles come under tension to support the weight of the head when it is tilted downwards in such a manner. This has implications in terms of viewing laptop screens and smaller sources of information such as tablets and smartphones. A computer screen should be set at a height where the top of the screen does not pass above the horizontal line of sight. Apart from ensuring that the screen can be read easily with the head in a more 'natural' relaxed position, screen users will never have to raise their eyes above the horizontal line of sight nor raise their heads so that they look upward, both of which increase the workload for the muscles concerned with managing and controlling this movement.

There is an exception to the rule with regard to computer screen height, and that relates to users who wear bifocal glasses. If they generally look through the bottom segment of their glasses when viewing the screen this will cause them to tilt their heads back if the screen is set at the levels suggested above. These individuals may need to have their screens lower than would be recommended for others so that they can keep their heads in a more natural position.

Having considered head position relative to a screen, thought also has to be given to any documents that might be used in conjunction with the screen. It is common practice to place documents on the desk surface to one side of the keyboard. This ensures that the worker has to tilt his head downward and to one side as they refer to the document (see Figure 1.6). To adopt a more appropriate head position, the individual should raise the documents off the desk surface. This could be achieved by placing the documents on a copy stand or document holder, assuming the documents are of a size that permits this. The document holder should be positioned alongside the screen and at about the same height. A better alternative is to use a document plinth which is similar in size and shape to a footrest. This is positioned in the space between the keyboard and screen and allows the documents to be presented directly in front of the individual in a tilted and raised position but without interfering with their view of the screen display.

Should an individual decide that they would like to work from a standing position, they should aim to establish the same relationship between keyboard height and elbow height by raising the desk to the correct level as discussed above. In addition, they should also ensure that the screen is presented at a height which enables them to work with their head in a comfortable position. See Figure 1.7.

FIGURE 1.6
A keyboard operator working with documents located on the desk surface.

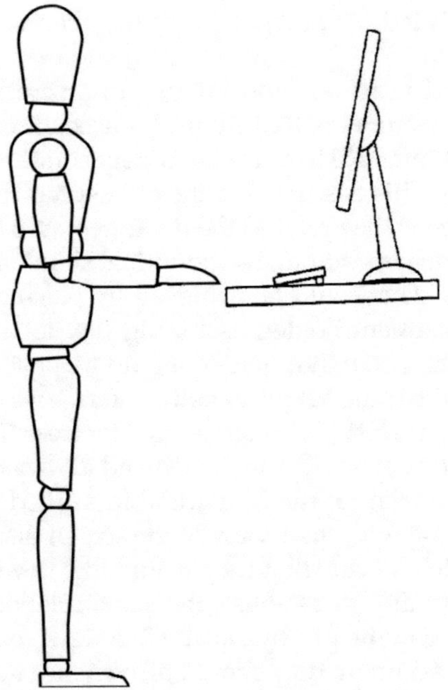

FIGURE 1.7
Posture when standing at a raised desk.

Some authors may suggest that the types of postures described in this chapter do not approximate actual working postures. They may suggest that they are too rigid, too 'idealised'. The fact is that the postures described above are based on what will impose the lowest loading on any part of the body active in the adoption and maintenance of working postures. What should be kept in mind is that the postures detailed in this chapter are a starting point. It is accepted that any posture, even one considered to be 'ideal', will become fatiguing if maintained for extended periods without interruption. Zemp et al. (2016) identified that subjects who reported back pain showed a clear trend towards a more static sitting behaviour when at work. Therefore, it could be concluded that office workers should move more during their working hours.

What is being advocated in this chapter is the provision of information to workers that will allow them to adopt postures known to be sound, and then allow them some flexibility around these postures. These postures are not expected to be adopted and maintained rigidly. Rather, the postures should have a fluidity about them so that individuals can alter their position within an acceptable range that still allows them to work comfortably.

1.5 Maintenance and Monitoring of Good Postures

Even if the workers appear to be fully committed to the concept of adopting suitable working postures and have made the effort to adjust their workstation equipment so that they can work comfortably, it is possible that over time the appropriate working practices will deteriorate. This could result from a number of factors, such as a new intake of untrained and/or inexperienced personnel, a change in the workstation furniture or equipment in use, a repositioning of equipment within the office or within a building, forgetfulness over time, trying to accommodate an illness or injury, or 'hot-desking' situations in which any number of individuals might use a given workstation.

An organisation should offer reminders at regular intervals about the need to adopt suitable working postures. Whether this is successful ought to be monitored carefully. Caution should be taken if any organisation is solely reliant on self-assessments, such as those that can be done on-line, as a means of gathering feedback. It cannot be assumed, even after thorough training, that all workers will employ the techniques reinforced during training or awareness sessions. It is not uncommon for individuals to complete self-assessment forms inaccurately because they do not appreciate that the postures they have adopted do not actually correlate with what has been suggested during training—which itself can sometimes take place using on-line facilities. Others may feel such time pressure from work demands that they pay fleeting attention to the questions on the checklist and complete it so that everything appears satisfactory and, as a result, they will not be 'bothered' by any further investigations. Although there is a place for self-assessments, particularly in large organisations, they should be used in conjunction with other monitoring procedures such as follow-ups by health and safety personnel, managers, or team leaders.

1.6 Summary

- If the task lends itself to an individual sitting down, a suitable chair should be provided to allow them to do so.
- Sitting offers advantages over standing in that people can take the weight off their legs, have greater stability for the upper body, and reduce energy expenditure and demands on the circulatory system.
- Standing desks provide opportunities for the person to change their posture.
- Prolonged standing can result in physiological changes, including an increase in heart rate.
- There is general agreement that people are not designed to sit for extended periods without interruption. Many individuals work in a manner that is likely to increase feelings of discomfort. Much of this is due to lack of awareness and training.
- The back consists of many parts that can be subject to general wear and tear, disease, the aging process, and abuse.
- Because the weight borne by the lumbar vertebrae is greater than that borne by either the cervical or thoracic vertebrae, the lumbar region is subject to higher levels of stress.
- The intervertebral discs are like cushions that sit between the vertebrae and act as shock absorbers. They also give flexibility to the spine.
- When an individual bends in any direction, the disc can become wedge-shaped and a bulge can form on one side.
- Bulges or ruptures can occur in the disc as a result of excessive stress, particularly imposed by bending.
- Protrusions caused by a prolapse can result in sciatica.
- The back muscles do the least amount of work when the body is upright and balanced. Once an individual leans forward, the

muscles contract to prevent them falling forward. The muscles also have to manage the movement. The farther forward one leans, the more active the back muscles become.

- Static muscle work is required to keep an individual in an upright sitting position.
- Static muscle work involves prolonged and uninterrupted contraction of the muscles.
- Dynamic muscle work involves contraction and relaxation of muscles, which achieves movement.
- Longer periods of rest are required to recover from static muscle work than from dynamic muscle work. Most tasks involve a combination of static and dynamic work.
- Static muscle work can result in discomfort quite quickly. More significant pain can be experienced if the individual does not take a break.
- The ligaments play a part in stabilising the individual in an upright position.
- Ligaments and muscles are susceptible to injury as a result of twisting and stretching, particularly over extended periods of time.
- The starting point for adopting suitable working postures is to ensure that the upper body is reasonably upright. This is not the same as suggesting that the upper body needs to be completely erect.
- Sitting in an erect posture is more likely to result in forward slump.
- Postures that result in operators leaning farther back than a reasonably upright position will move them away from the desk and keyboard, resulting in them having to extend their arms. This is likely to be fatiguing.
- The mouse and keyboard should be located close to the leading edge of the desk. A 10 cm gap in front of the keyboard is sufficient to allow users to rest their wrists in between bouts of keying.
- Assuming a fixed-height desk is in use, a chair should be adjusted so the individual's elbows are level with the home row of the keyboard.
- If the person's feet are not firmly on the floor once the chair has been adjusted, the person should be provided with a footrest.
- A footrest should not be used by anyone whose feet naturally rest securely on the floor.
- Working without a footrest when the feet do not touch the floor when in a seated position will create problems, such as the leading edge of the chair compressing the backs of the thighs due to the weight of the legs. This will cause discomfort.

- If adjustable-height desks are in use, users should position the chair so that their feet are firmly on the floor, and then move the desk into a position where they are able to operate the keyboard when it is at the same height as their elbows.
- Once the user's sitting height has been established, the backrest needs to be altered. The lumbar support on the backrest should be positioned so that it is lined up with the small of their back.
- When operating the keyboard, the user should not rest the wrists on the desk. They should hover over the keyboard relying on the larger arm muscles to move the limbs into position rather than having to overextend the fingers to depress the keys.
- If armrests on chairs prevent users from sitting as close to the leading edge of the desk as they might wish, the armrests should be removed.
- Head and neck posture are determined by the visual requirements of the task.
- The computer screen should be positioned so that the top of the screen does not move above the user's horizontal line of sight.
- Users wearing bifocal glasses may prefer that the screen is presented at a lower level so that they do not have to tilt their heads upward and back in order to look through the bottom of their glasses when viewing the display.
- Documents should not be placed on the surface of the desk and referred to when the individual is operating the keyboard. People should be encouraged to use a document holder or copy stand to present the document at the same height and distance as the screen or on a document plinth in front of the screen.
- Once appropriate working practices are established, it is likely that they will deteriorate over time for a number of reasons. Organisations need to offer reminders at regular intervals about the need to adopt and maintain suitable working postures.
- Caution should be taken when using self-assessments because employees do not always complete these accurately, and they do not always correlate the information they have gathered online with their own particular workstation.

References

Anderson, D. G. and Tannoury, C. 2005. Review molecular pathogenic factors in symptomatic disc degeneration. *The Spine Journal* 5 (6 Suppl), 260S–266S.

Barbieri, D. F., Mathiassen, S. E., Srinivasan, D., Dos, S., Wilian, M., Inoue, R. S., Siqueira, A., Almeida G., Nogueira, H. C., and Oliveira, A. B. 2017. Sit–stand tables with semi-automated position changes: A new interactive approach for reducing sitting in office work. *Occupational Ergonomics and Human Factors* 5 (1), 39–46.

Bridger, R. S. 2003. Introduction to ergonomics. 2nd ed. London: Taylor and Francis.

McKeown, C. 2011. *Ergonomics in Action: A Practical Guide for the Workplace*. Abingdon, UK: Routledge.

Pheasant, S. and Haslegrave, C. 2006. *Bodyspace. Anthropometry, Ergonomics and the Design of Work*. Boca Raton, FL: CRC Press.

Watkins, J. 2009. Structure and function of the musculoskeletal system. 2nd edition. Human Kinetics Champaign Il.

Zemp, R., Fliesser, M., Wippert, P., Taylor, W. R., and Lorenzetti, S. 2016. Occupational sitting behaviour and its relationship with back pain—A pilot study. *Applied Ergonomics* 56, 84–91.

Zhang, Y., Guo, T., Guo, X., and Wu, S. 2009. Clinical diagnosis for discogenic low back pain. *International Journal of Biological Sciences* 5 (7), 647–658.

2

The Design of Workstation Furniture

2.1 Introduction

When considering the design of workstations, the initial focus will generally be on the individual pieces of equipment as standalone items. However, the interaction of items when used in combination is also very important. This chapter will consider the design, layout, and use of desks, chairs, and accessories such as footrests, wrist rests, document holders and plinths, headsets, and reading slopes, and the impact of one on another when used in conjunction.

Wireless technology has changed the way in which people are able to work. They are not bound to their desks as they formerly were. People can perform computer-based operations virtually anywhere within and outside the office. Some of these places could be considered 'atypical' offices, and these will be discussed in Chapter 5. The majority of computer users, however, still sit at dedicated workstations within the office to perform their work.

2.2 Desks

Desks used to be such an uncomplicated feature in an office: simple rectangular shapes, sometimes with drawers attached to the undersurface. However, as technology has changed, desk designers and manufacturers have been working assiduously to keep up with the demands created by the need to support screens, keyboards, mice, telephones, printers, scanners, and other devices on the desk surface, and to accommodate increasing numbers of individuals in an office without requiring additional floor space. Desks are now contoured and shaped, and can be used individually, in mirrored pairs, or in 'pods'—large groups of desks joined together making up a team's workspace. They can be fixed-height, partially height-adjustable, or fully height-adjustable, and some can be converted to standing desks. They can have drawers attached to the undersurface or have independent pedestal drawers.

They offer a wide range of cable management facilities. Any brochure that accompanies these desks will advise the reader that their designs meet a range of design specifications and legal requirements. Yet many of them will fail from the perspective of usability and functionality once introduced into a real working environment. This is usually because they do not include some simple, but fundamental, design features that take specific account of the work that will be performed at the intended workstation.

2.2.1 Desk Height

There was a time when adjustable desks were significantly more expensive than non-adjustable desks, which restricted the number that would have been introduced into any office. However, that is not necessarily the case now. Other factors typically place restrictions on the use of such desks. In the main, office managers and designers tend to favour a uniform look in an office, so every individual has the same workstation type and layout. Pod-style arrangements, where several workstations are conjoined, do not lend themselves to individuals setting the desks to different heights. As a consequence, office workers tend to be provided with fixed-height desks.

There is absolutely nothing wrong with using a fixed-height desk—one that cannot be altered for height—as long as a suitable height-adjustable chair is made available and a footrest is provided, if required. The provision of a suitable chair and footrest becomes particularly important in 'hot-desking' situations where more than one individual might use a desk during the course of a working day. This is common in offices that operate multiple shifts offering 24-hour cover, such as call centres, where three individuals may use the same workstation during a 24-hour period. It is also common in offices where only a small percentage of the employees would be expected to be in the office on any one day, such as travelling sales staff. In this case, it would not make business sense to provide and maintain a workstation for each employee if only a small number of the workstations would be used at any one time. Many organisations combat the influx of personnel on core days, when more bodies than desks are present, by providing break-out areas and no-frills desking areas that provide a power point and just enough space for a laptop.

Fixed-height desks tend to be designed so that they offer a working height of 720–740 mm (28¼″–29″). This is an industry 'standard' height, although some fixed-height desks may be in use that have a lower working height than this (these tend to be older desks), and other desks may have a higher working height, possibly as much as 750–760 mm (29.5″–30″), but these are unusual.

Any organisation will find that using a standard height desk will suit the majority of the individuals in their offices. The only exceptions to this will be taller individuals, usually tall men, and possibly people with specific disabilities. It might be assumed that small individuals, such as petite females, may be disadvantaged by using such a desk, but this is not necessarily the case. Small individuals can always use adjustable chairs to raise themselves to an appropriate height relative to the desk surface and keyboard (as described in Chapter 1), and then support their feet on a footrest.

Companies should be alert to the fact that some of the smaller individuals may work on the erroneous assumption that their feet must touch the floor, and, as a consequence, they may lower their seats. As they reduce the seat height, the height of the desk and keyboard will, in effect, rise relative to their seated position. This will force users to raise their arms to operate the keyboard. This posture will be maintained throughout the period that they remain at their desks, and this will require uninterrupted static muscle work. This will be fatiguing and often leads to discomfort between the shoulders. To judge whether a person is working at too low a level relative to the desk or keyboard, stand behind them when they are working and observe whether their elbows and upper arms are sticking out from their body sides (see Figure 2.1). The posture that should be adopted when using a fixed-height desk is outlined in detail in Chapter 1.

If a taller individual, usually in excess of 183 cm (6″) in height, finds that a standard desk is too low, a number of options are available, including the provision of a height-adjustable desk.

Height adjustment features come in a variety of forms. Some manufacturers have incorporated telescopic legs into their desk designs so that the legs can be extended to present the desk surface at a suitable height. This

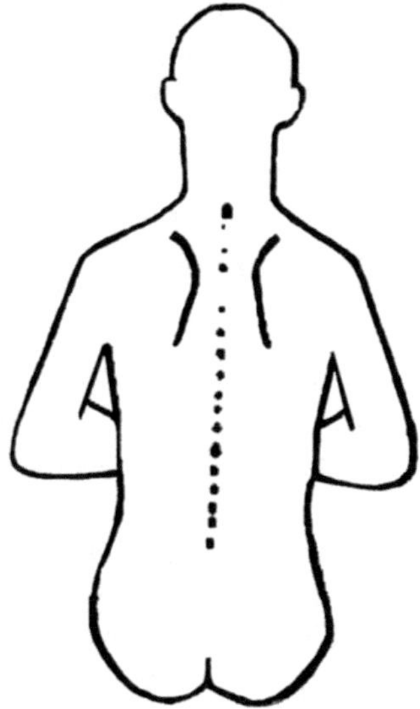

FIGURE 2.1
The likely upper limb posture of someone sitting at too low a level relative to their desk and keyboard.

feature has a few minor drawbacks: the adjustment can be done safely only without anything on the desk surface, which makes it more difficult and time-consuming to adjust once the desk is in use; and care has to be taken to ensure that each leg is set at exactly the same height so that the desk surface is level. Some manufacturers may suggest that the legs of their models can be adjusted for height, but, in actual fact, all they offer are stabilising feet that are adjusted to accommodate an uneven floor surface. They may, in reality, be capable of changing the height of the work surface by only a few millimetres or a centimetre.

Alternative height-adjustment mechanisms include crank handles located on the surface of the desk, directly over the leg area, and electrically powered control panels embedded in the desk surface. With the crank handle method, a chain, similar to a bicycle chain, is located in one of the desk legs, and as the handle is rotated the chain moves, which raises or lowers the desk.

Electrically powered height-adjustable surfaces often have a small panel on the desk surface or leading edge that presents 'up' and 'down' buttons that, when pressed, allow the surface to move up and down smoothly and slowly. This feature is particularly useful for individuals who might require different working heights for different tasks, such as when writing and then using the keyboard, or for those who have a disability that might not allow them to operate a crank handle easily.

Some adjustable desks provide a split surface—for example, split horizontally to provide front and rear sections—where each section of the surface can be adjusted independently. This can be a useful feature because it allows the screen, which would be placed on the rear section, to be positioned at a suitable height relative to the seated user. However, this design will only work successfully if the front section of the desk can comfortably accommodate the keyboard, mouse, etc. It should be noted that a split surface desk is different to a keyboard tray which is added separately to the undersurface of the desk. See comment below on using these add-ons.

When using adjustable work surfaces, users should, ideally, lower their chairs so that their feet are flat on the floor. The desk surface should then be lowered so that the individual is able to adopt the posture outlined in Chapter 1, with the elbows level with the home row of the keyboard.

Care has to be taken that an individual's sitting position alongside the desk is not compromised by some feature on the undersurface of the desk. An example of this would be a poorly positioned pull-out keyboard tray designed to be stowed under the leading edge of the desk when not in use. These commonly come in contact with the tops of a user's thighs when they try to raise the chair and restrict the height at which they can sit. Sometimes, this can result in the user sitting too low relative to the desk or screen. There should be a clear space of at least 650 mm (25.5″) between the floor and the undersurface of the desk.

2.2.2 Work Surface Design

One of the most important features of the desk is the surface area available. This will be a major determinant of how individuals can arrange their equipment. Ideally, they should be able to arrange the screen and keyboard so that when they are in use, they are positioned one in front of the other, directly in front of the user. As a general rule, the depth of the desk should allow the screen to be positioned about fingertip distance away from the user when they are in their 'at work' position. The screen should not be positioned to one side of the desk so that the user has to look to the side each time they refer to the display. This was a common feature in offices when screens had become large and bulky but were still placed on older rectangular desks that were not deep enough to allow the screen to be positioned in front of the user. This also occurred in situations where individuals had many documents to work with and moved their screens and keyboards to one side of the desk to accommodate the documents. If the screen is positioned to one side, users will have to rotate their heads each time they refer to the screen and again as they look back at their documents or keyboard. This layout can also cause twisting of the upper body. Delleman (2004) has detailed a study that showed that visual display unit (VDU) operators who twisted their necks beyond 20° showed a higher incidence of neck and shoulder disorders.

A screen should only be positioned on one side of the desk if the user is able to reposition the chair so that they can align themselves in front of the equipment when it is in use. The only type of user who could possibly justify working with the screen to one side of the seated position would be a copy typist who simply reads printed or written text as they type and does not refer to the screen throughout the keying operation. Even copy typists, however, should sit in front of the screen if interacting with it, such as when editing previously typed text.

Given the fact that most individuals in an office have to work with both computers and documents, manufacturers have provided desks that offer increased surface area. They have accomplished this by offering cockpit-style surfaces, L-shaped surfaces, or 'wave' surfaces (see Figure 2.2). Many of these desks do not take up any more floor space than rectangular desks did; therefore, the same number of people can be accommodated in an office without resulting in cramped conditions. The key issue with all of the desks being offered, regardless of their surface shape, is whether they can allow the appropriate positioning of the screen. This is easier to achieve with single flat screens.

Using the available surface becomes more problematic if more than one screen is in use. Many desks can accommodate two screens, particularly flat screens. But once a user has to refer to more than two screens, a specifically-designed desk and layout should be provided.

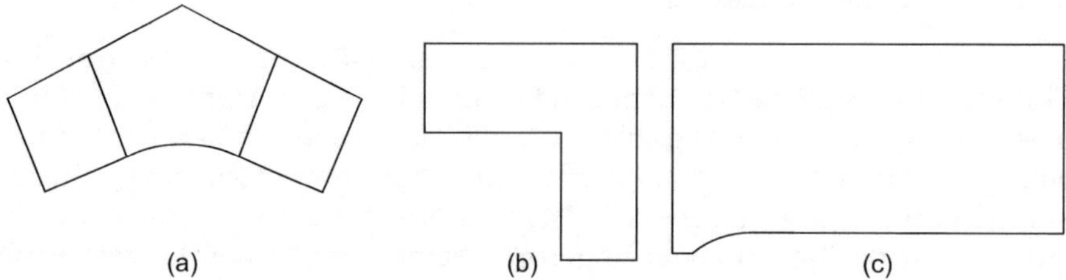

FIGURE 2.2
Desk surface configurations: (a) a cockpit-style surface, (b) an L-shaped surface, and (c) a 'wave' surface.

The desk surface also needs to provide sufficient space to accommodate the mouse, documents, and copy stand, a telephone, if in use, and any other equipment that will be utilised at the workstation. Final purchasing decisions need to be made with care, preferably following user trials. Often the most creative-looking designs fail in terms of usability when they are introduced into a real working environment. A prime example of this would be the split-level computer table primarily aimed at the home market. It is intended to represent a compact means of supporting a screen, keyboard, mouse, and other devices. Often, the lower surface, intended for the keyboard, offers such a small surface area that it can accommodate only the keyboard and does not leave sufficient space to move the mouse. Often, it cannot even provide sufficient space for users to rest their forearms in between bouts of keying. To compound the problem, sometimes there can be a secondary shelf under the main work surface to support the printer. This reduces the leg space available to the seated user.

Other popular 'shaped' designs also have to be selected with care. The L-shaped work surface and 'wave' surface shown in Figure 2.2 can create problems if not utilised effectively. For instance, Chapter 1 advised that people should be able to sit close to the leading edge of their workstation when operating a keyboard. As Figure 2.3 shows, if a user sits on a chair with armrests, they will frequently find that the armrests come in contact with the leading edges of the L-shaped desk where they form a 90° angle, and this

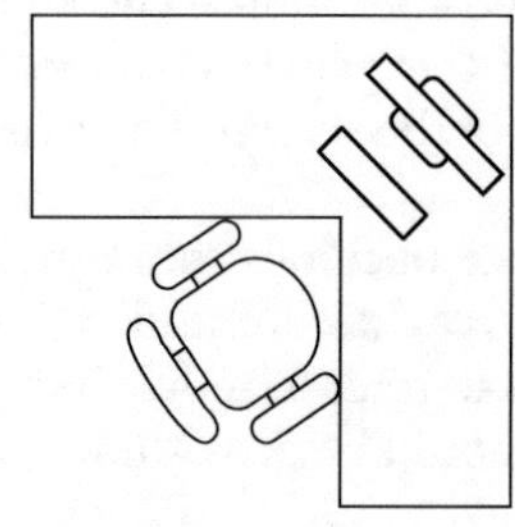

FIGURE 2.3
Possible problems caused by using a chair with armrests alongside an L-shaped desk.

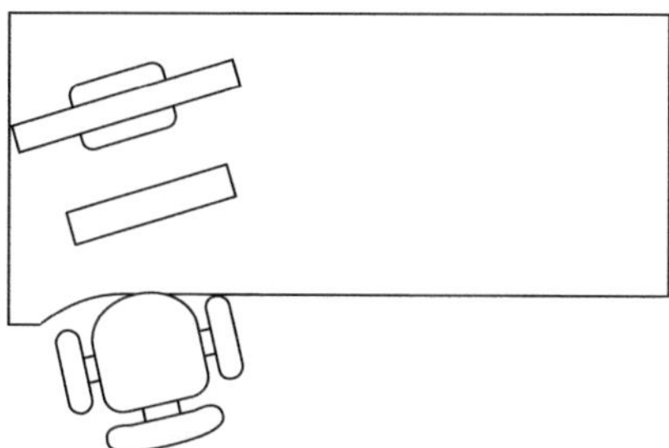

FIGURE 2.4
Possible problems caused by using a chair with armrests alongside a 'wave' effect surface.

prevents them from sitting as close to the desk as they might wish. A similar problem could be experienced with the contoured desk. As the user turns toward the screen, which involves rotating the chair, as shown in Figure 2.4, the armrest can come in contact with the side of the desk and push the user off-centre in relation to the screen. Users might find that they have to compensate by leaning slightly to one side to feel they are in line with the equipment, or they might have to move the equipment out of the designated area on the desk to allow them to sit more comfortably, resulting in poor placement of the equipment.

Desks with asymmetric surfaces are often used in pairs or pod arrangements and are provided as mirror images, such that one is referred to as a right-handed desk and the other as a left-handed desk. This means that the major proportion of the user's workspace will be located to either the left or right side (see Figure 2.5). Some individuals will complain that this limits flexibility over how they use the workspace. When desks are used in back-to-back arrangements, usually divided by a low-level partition, this limits how far back equipment and other items can be moved. As a result, screens can end up being too close to the user, particularly if they are supported on

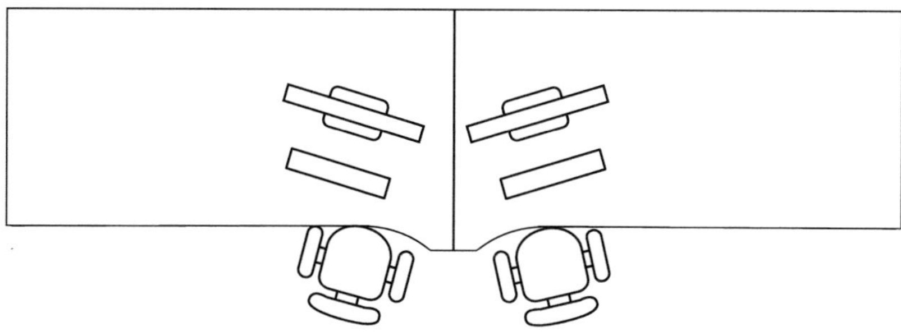

FIGURE 2.5
Asymmetric surfaces used in a paired formation, resulting in workspace to the left or right of user.

a screen arm which can be bulky and which would, under normal circumstances, overhang the rear of the desk.

The impact of stylised desks can only really be determined through trials of the furniture in the real workplace prior to purchasing. Most desk suppliers will agree to such a suggestion. The subject of trials is discussed in Chapter 3.

One other feature of the work surface that deserves mention is the leading edge. It was common for older desks to have 90° angle edges. This is still the case with many less expensive models currently provided by suppliers. The concern with this design feature is that if individuals choose to rest their wrists or forearms on the sharp edge of these desks, this can compress the tissues of the wrist and forearm and can, over time, contribute to the development of an upper limb disorder. Desks with rounded or 'bull-nosed' edges are more appropriate.

2.2.3 Work Surface Layout

Having selected the appropriate desks for the office, effort should be made to ensure that the layout is appropriate. A badly planned work surface layout can result in unnecessary overreaching, leaning, and twisting. The aim should be for the seated worker to be able to reach all frequently-used items easily. This is important because the seated worker is virtually fixed in place and cannot move towards an object in the same way that a standing worker can.

When considering the area in which an individual can reach, consideration is given to the 'zone of convenient reach' (ZCR) and the 'normal work area' (NWA). The ZCR is sometimes described as the secondary work area and covers an area that can be swept easily by the fully extended arm as it is moved sideways, upward, and forward. The ZCR for a seated worker is less than that for a standing worker (Sengupta and Das 2000). Within the ZCR is a smaller area, or NWA. Both are shown in Figure 2.6.

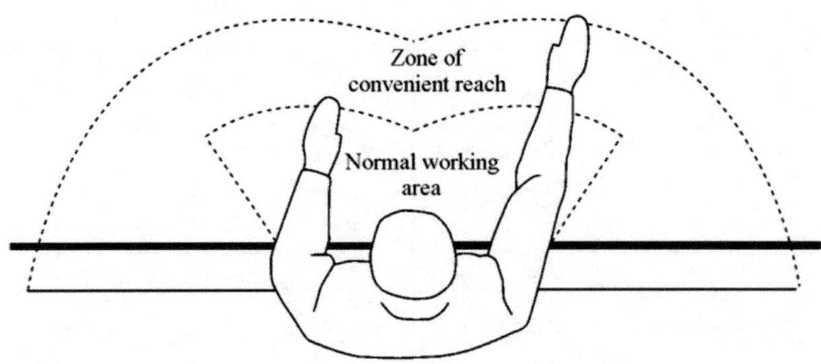

FIGURE 2.6
The zone of convenient reach and the normal work area. (From McKeown, C., *Ergonomics in Action: A Practical Guide for the Workplace*, CRC Press, Boca Raton, FL, 2011.)

The NWA is usually referred to as the 'primary work area' and represents an individual's optimum reaching distance. It covers the area that can be comfortably swept by the forearm when it forms a 90° angle at the elbow with the upper arm, which hangs naturally by the side of the body. Any activities that have to be performed for extended periods, or repeatedly, should be carried out within the NWA or primary work area. Any tasks that are performed on an irregular basis, or for short periods, can be carried out in the ZCR or secondary work area. This principle should be applied when arranging equipment and any other items on the desk surface. Those items used regularly—for example, the mouse, the keyboard, and perhaps the telephone—should be located within the NWA, and those items used infrequently, such as reference folders, directories, and perhaps the telephone, can be located in the more distant areas of the desk.

Some individuals may find that they have so much documentation to refer to when they work that the surface area available is insufficient. As a consequence, they may choose to store the overflow from their desk on the surrounding floor. This is likely to lead to reaching to the documents from the seated position, which will cause twisting and stooping. This is considered particularly stressful for the lower back and should be discouraged. If someone has a valid reason for having so many files to work with that they are driven to place them on the floor, this probably warrants a larger workstation or, alternatively, an additional mobile surface, such as a trolley, to accommodate the overflow. Ideally, nothing should be stored on the floor, as it will only encourage bending and stooping, as well as present a tripping hazard.

2.2.4 Undersurface Features

The design of the undersurface of the desk is just as important as the work surface, because it can have an impact on comfort levels. An individual should have sufficient leg clearance from front to back and from side to side to be able to move the legs freely and change position easily without hindrance. It is generally recommended that someone sitting at a desk should be provided with a minimum clearance of 600 mm (23.5″) from front to back under the desk. A clearance of at least 600 mm (23.5″) from left to right relative to the seated position is also recommended; although, ideally, it should be 1000 mm (39.5″) in this area. Consideration should also be given to the location of features such as modesty panels toward the rear of the desk, desk frames, and cantilever legs which can become an obstacle at floor level. In situations where desktop computers are in use, the processors are often placed in the immediate foot area under the desk surface. They are frequently located there by individuals who do not even work at the workstations in question, such as IT personnel.

Drawers can also limit the user's freedom to move. Pedestal drawers were considered to offer a degree of liberation to users because they were intended to allow them to push the drawers to an unobtrusive area. However, the reality is that most pedestal drawers are located under the desk surface in exactly the same place as fixed drawer units would have been.

Cable management should also be given thorough consideration. Most desks offer a facility for storing or controlling cables, but many office workers fail to utilise these effectively. Part of the problem is the fact that desks and their users get moved around in offices regularly, and it can be time-consuming to remove cables from their storage points in the desks during the transfer and then to reinstate them. As a result, cables lie on the floor area around the feet or around the desk, representing a risk to the user and passers-by. Additional consideration needs to be given to the impact on cables if the desk is intended to be raised upwards so the user can stand. Cable length and cable entrapment become important in this situation.

2.2.5 Sit–Stand Desks

The use of sit–stand desks has increased in popularity recently. It is generally believed by office workers that if they can change the height of their desks and alter their posture throughout the day so that they alternate between periods of sitting and periods of standing, they are likely to be more comfortable. Evidence provided by Wilks et al. (2006) suggests that alternating between sitting and standing will reduce musculoskeletal discomfort and improve worker well-being; this is supported by Pronk et al. (2011) and Karakolis and Callaghan (2014) who concluded that use of sit–stand workstations is associated with reduced discomfort. Robertson et al. (2013) also demonstrated that those who took advantage of the standing option in a sit–stand workstation improved their work performance.

As a means to explain the reduction in discomfort experienced by alternating from sitting to standing, Marshall et al. (2011) and Nelson-Wong and Callaghan (2010) have suggested that a standing workstation may afford users more opportunity to employ a range of postures which will result in more frequent muscle activity whilst trying to support these postures. In addition, Karakolis et al. (2016) reported that benefits were gained when returning to the seated position from standing, as it allowed for variations on the original sitting position to be adopted on each occasion that the user sat down again.

Despite the apparent positive aspects of using sit–stand desks, there also seem to be some possible drawbacks. Although performing computer work when seated was associated with more non-neutral shoulder postures and greater shoulder muscle activity in a study by Lin et al. (2017), doing the same work at a standing desk was found to result in greater deviation of the wrist. Lin et al. were also able to establish that although users on both standard seated desks and standing desks reported similar overall discomfort levels within the first 10 minutes of starting work, they reported twice as much discomfort while standing than sitting after 45 minutes, with most discomfort reported in the lower back. This might go some way to explaining why a number of users, who have been provided with sit–stand workstations, ultimately fail to routinely adjust the workstation height so they can stand after a few months of use.

Given that using a standing desk prevents the user from employing all of the principles passed on to them during training in the use of a 'standard' desk, it is important that they are provided with instruction in the use of their standing desk and guidance on what is considered suitable working practice. Lin et al. (2016) have pointed out that many users chose workstation set ups for a sit–stand workstation that do not completely conform to recommended guidelines.

2.3 Partitions

With the advent of open-plan offices, partitions have been employed to offer a degree of privacy, acoustic control, and to provide demarcation between departments or teams. Before any decisions are made about the form the partitioning should take, full consideration should be given to the task demands of each individual enclosed by a partition and the psycho-social issues relating to their use.

The immediate impact of partitions can be isolation. This has been referred to as an 'unintended influence'. Although it might be assumed that contact between people has a big effect on the cohesiveness of a group, and partitions act as physical constraints that impede the ease with which people can interact, De Croon et al. (2005) suggest there is little evidence to support the view that open-plan offices result in a worsening of interpersonal relations. However, they did find that there is strong evidence to suggest that open-plan workplaces reduce psychological privacy, and job satisfaction.

Although there could be drawbacks in some situations, partitions may be suitable for a job requiring confidentiality or for individuals who use the telephone frequently, which may be distracting for surrounding workers. In addition, partitions clearly mark out an individual's personal space, which cannot be encroached on by other individuals sitting nearby.

The success of partitions lies in identifying whether they are a real requirement of the work or conditions, and then deciding what form they should take. They can be high-level so that a seated individual cannot be seen by anyone near them, or they can be eye-level, which allows seated workers to see other people sitting around them and to communicate, albeit with some difficulty, when necessary.

2.4 Chairs

Chairs have to be selected after a thorough review and testing of what is available on the market. Because most desks currently in use in offices cannot be adjusted for height, an individual will only be able to adopt a suitable, fully-supported working posture by adjusting the chair. Individuals performing a screen-based operation at work, or even a simple pen-and-paper task, should not be using non-adjustable chairs, unless their task lasts for only a few minutes at a time and will not be repeated at regular intervals.

There have been attempts at providing alternative forms of task seating in the office. Wittig (2000) compared varying designs such as pendulous chairs, sitting balls, kneeling chairs and raised chairs against a standard office chair. No significant differences were found in the muscles that straighten the back or the extent to which the seated user leaned forward. Interestingly, findings relating to the subject's perception of the benefits to be gained by using the alternative seating indicated that it should not be used as a permanent seat in place of a standard office chair. Van Dieën et al. (2001) analysed the effect of three dynamic office chairs on trunk movement, trunk extension to sit more upright, and spinal shrinkage. The results showed trunk movement and trunk extension were influenced more by the specific task being performed at the desk than by the chair type. This would underline the importance of ensuring the work surface is set up correctly so that equipment is in a suitable position relative to the user and everything is within easy reach.

As a minimum, the chair should be adjustable for height. The backrest or independent internal lumbar support should be designed so that it can be repositioned relative to the seated user; alternatively, the backrest should be constructed of a material that moulds around the individual. It should have five prongs on the base with casters (unless the floor covering makes this unsuitable), and the padding should be sufficient to prevent the user from coming in contact with or being aware of hard edges on the shell of the seat.

2.4.1 Seat Height Adjustment

The range of adjustment offered by office chairs varies from one model to another. Ideally, a seat should be capable of adjusting from around 380 mm (15″) to 530 mm (21″) above the floor. Typically, office chairs do not offer the full range that might be considered 'ideal', but they are still likely to accommodate many potential users. Only tall or very small individuals might find that a chair does not move enough to accommodate them. This can be dealt with easily by simply approaching suppliers and requesting a chair on trial that offers a greater range of height adjustment than the 'standard' chair.

2.4.2 Backrest Adjustment

The backrests of chairs are varied in style and many are now quite sophisticated in design. Each style of backrest has its merits and a number have drawbacks. It is important to be aware before making any purchasing decisions what the pros and cons might be for each design.

The aim of the backrest should be to offer support to a significant proportion of the back. As a minimum, this support should be available from around the small of the back to just below shoulder level. As the small of the back is usually concave when an individual adopts a suitable sitting position, the backrest should be shaped so that its lower section, the lumbar support, fits neatly into the small of the back. To execute this efficiently, the lumbar support should be capable of being moved relative to the seated user, or the material of the backrest should be capable of moulding around and supporting the individual. The lumbar support can be moved in a number of ways depending on the design of the chair.

The most usual method of moving the lumbar support into position is to move the whole of the backrest up or down. Alternate methods include sliding a lumbar roll up and down inside the backrest. Some chairs have contoured backrests with distinct lumbar areas, but the whole backrest is fixed to the seat offering no form of adjustment. Given that users will come in different shapes and sizes, it will be rather hit-and-miss whether, once they sit down, their lumbar region lines up properly with the immovable curve of the seat. If it does not line up, the individual will be forced to adopt a posture dictated by the design of the seat. If such a mismatch is evident, the user should not be expected to sit in the seat.

In addition to offering a feature to change the position of the lumbar support, backrests should be capable of tilting. This will enable users to vary their sitting positions, within an acceptable range, throughout the day.

Some backrests have what is referred to as a synchronised or rocker motion. This means that the backrest moves backwards and forwards as the individual moves and does not actually lock in a predetermined position. Although this is advertised by suppliers as being beneficial, many users find it irritating because they want support from a solid, non-moving surface. This should be taken into account when making selections.

A number of seats have been designed so that as the backrest is tilted, the seat tilts also. Although there may be merit in offering the user greater choice over how they sit, the design should not dictate a posture to the individual. This can occur if the seat and back move in unison so that the position of one dictates the position of the other. If a chair is intended to provide this combination of adjustability, it should be offered in a way that allows the user to alter the seat and backrest independently of each other.

Some users believe that they require high backrests and some request the provision of seats with head supports. A higher backrest does not necessarily mean better support. Some users find that higher backrests, which cover the area from the small of the back to the nape of the neck, actually interfere with the movement of the shoulders and arms as they work. Only a full trial of a seat will indicate whether this is likely to happen.

The use of a seat with a head support has to be questioned in a work setting. To use this feature effectively, the individual has to deliberately lean the head back onto the rest. This is likely to be set in a position from which the user cannot hope to view the screen adequately and comfortably. The head support is effective only when it is being used as a headrest, when the individual is not actively working with their computer.

Some smaller users may find it hard to use the backrest effectively. This usually results from the seat being too large. If the small user sits back in the seat, the leading edge of the chair will protrude into the backs of the knees, making them uncomfortable. As a consequence, small users tend to perch on the front of their seats, losing all support from the back of their chair. These individuals should be provided with alternative chairs that have reduced dimensions, thereby allowing them to sit back in their seats.

2.4.3 Seat Slide

Having a facility to extend the length of the seat is a common feature in many office chairs. In essence, the seat pan, on which the user sits, is capable of sliding backwards and forwards so that the sitting area can be increased or reduced in depth. This is a useful feature for taller users with longer legs who require additional support and for smaller users who would normally find that the seat is too long and protrudes into the backs of their knees. The only drawback of this facility is that the control mechanism is usually well hidden, and many users are unaware that the facility is available, and, therefore, do not take full advantage of its presence.

2.4.4 Armrests

Armrests should be considered a serious topic for discussion. Many workers in an office have a number of misconceptions regarding the presence, or absence, of armrests. Typically, armrests were considered to reflect the rank of an individual within the hierarchy of a business. The higher up the chain an individual was, the more likely they were to have armrests on their chair—which itself would typically be bigger, more sophisticated, and more expensive than anyone else's chair.

A number of users believe that it is a legal requirement to have armrests on the chair. This is not the case. Other users think that they are less likely to develop an upper limb disorder if they have armrests. This is also untrue. Armrests, particularly poorly designed and badly positioned armrests, can create problems for users if they are permanently attached to the chair. The main problem is that some armrests prevent users from sitting as close to the leading edge of the desk as they might wish. As a consequence, users are presented with a greater reaching distance than they might prefer between them and the keyboard or mouse. To overcome this hurdle, they either extend their arms forward, which increases the workload for the arms, or they sit on the front edge of their seats, thereby losing all support from the backrest.

Some users think they can avoid this problem by lowering their seats, which will enable the armrests to move under the desk surface as they pull the seat close to the desk edge. Unfortunately, this tactic will only create new problems. As the chair is lowered, the height of the work surface, keyboard, and mouse rises relative to the user's seated position. The user will then be forced to raise the shoulders and arms in order to reach the keyboard and mouse. Static muscle work will be required to hold the arms and shoulders in the raised position, and this is extremely fatiguing and likely to result in discomfort (Pheasant and Haslegrave 2006). The individual is unlikely to change their position until they stand up and leave the workstation, which might mean that some continue to work in this irregular posture for several hours at a time without interruption.

Some armrests take into account the user's need to be able to sit close to the work surface. Some are adjustable for height, which allows them to be lowered to get them out of the way if required, and some can be adjusted in width, which allows larger users to be accommodated. A number of armrests have been reduced in overall length so that their upper supporting surface does not extend the full length of the seat, which would normally result in them coming in contact with the desk edge as soon as the chair is moved towards it. Other armrests can be rotated so that they move from extending forward to extending backward and are, in effect, out of the way altogether.

If armrests are present, they should be capable of being used in a way that offers support to the user's forearms, but does not interfere with the way the user works. The user should be able to move their arms freely as they reach for objects and as they operate the keyboard and mouse.

Should any individual be working with a chair that has armrests that are proving to be a problem, they can usually be removed easily. Most often, the armrests are attached to the main frame of the seat as secondary parts and can be removed with the aid of a spanner or Allen wrench or Allen key.

2.4.5 Adjustment Mechanisms

The key feature of any adjustment mechanism should be that it is easy to use from the seated position. Any adjustment that requires the user to stand up to operate the control is not appropriate for use. If controls are complex, users are unlikely to take the time to familiarise themselves with the controls and will not use them effectively, if at all. This statement is true of the chair's whole design. If a chair is very complex, users often fail to utilise it to its maximum capacity and often fail to use any of the functions properly.

The one adjustment mechanism that is least likely to be used on a chair is the tension adjustment found under the seat itself. The intention behind this particular control is to change the setting of the seat so that when a big, heavy user tries to lean back, there is sufficient resistance in the movement of the backrest to prevent them from feeling they are going to fall off the seat. When a smaller, lighter user sits in the same seat, they should change the tension setting so that less effort is required to alter the tilt on the backrest as they lean on it. Many users are unaware that such an adjustment exists.

Diagrams on each adjustment feature that offer intuitive explanations for the use of that adjustment are helpful. Training before using a chair is an absolute necessity. It should not be assumed that users already know how to use an adjustable chair and know what sort of posture to adopt. In addition, complete reliance upon on-line training packages as a means of illustrating how to use workstation furniture is misguided. Many users are unable, or unwilling, to transfer what they see by way of a demonstration into their own working position. Ensuring that users adopt suitable working postures can only be achieved by observing each of them briefly at their own workstations and offering them advice, if needed. If this is to be effective, advisers or assessors should have a good working knowledge of the equipment in use so that they can provide clear instructions, and demonstrations if required, on how to make alterations. Given that workers usually exceed recommendations regarding the maximum time they should remain working in a sitting position (Netten et al. 2011; Ryan et al. 2011; Goossens et al. 2012), it is important that even when users feel comfortable in their seat, they should understand the importance of leaving their workstation at regular intervals. Prolonged sitting can result in an increased risk of developing health problems (Chau et al. 2010; Healy et al. 2013).

2.5 Accessories

There are a number of items that are usually categorised as 'accessories' in office supply catalogues. These include footrests, wrist rests, document holders or copy stands, screen risers, telephone headsets, and reading slopes.

2.5.1 Footrests

As indicated in Chapter 1, if a user's feet do not touch the floor once they have altered the chair so that they are sitting at a suitable height relative to the desk or keyboard, they should be provided with a footrest. As the footrest is taking the place of the floor, it needs to offer a number of characteristics similar to the floor. It needs a large surface area so that it can comfortably support both feet without them tending to slip off. The surface needs to be covered in a non-slip material.

Ideally, as users have varying statures and thus have different leg lengths, a footrest should be adjustable for height and preferably tilt. If a user has a desk that is divided into two distinct work areas, which commonly happens with an L-shaped arrangement, they should be provided with two footrests—one for use under each section of the desk. Users who find that their feet do touch the floor once they have adjusted their chairs should not use footrests for the reasons identified in Chapter 1.

On some occasions, the footwear worn by the individual will have an impact on whether they need a footrest or not. As a consequence, a user's requirements will vary depending on whether they wear trainers/sneakers one day and heels the next.

2.5.2 Wrist Rests

Wrist rests are long sections of gel-filled fabric intended to lie in front of the keyboard. The biggest mistake people make is to assume that they should lean their wrists on these rests while they are keying. These rests should only be used when the user is not keying and should only be employed when the user is simply resting the hands. If individuals adopt the postures described in Chapter 1, they will find that their hands 'hover' over the keyboard. When this is the case, they can rely on the larger muscles in their arms to reposition their hands relative to particular keys. If, however, the user leans on the wrist rest while attempting to operate the keyboard, they will effectively be 'anchored' in place and will only be able to reach individual keys by extending their fingers. This is not a suitable way to work. Should an organisation provide wrist rests to users, they must also provide appropriate accompanying information so that users are not under the illusion that they must lean on the wrist rest while operating the keyboard.

2.5.3 Document Holders

Research suggests that there is a definite threshold at which there is a risk of developing musculoskeletal disorders as a result of repeated head and neck movement (Szeto et al. 2005). Neck posture has been found to be a risk factor for neck pain (Delleman 2004). It has been established (Thorn 2007) that those who experience neck pain at work are likely to move their head and neck more frequently and do so with reduced rest periods.

Typically, many users place their source material on the desk surface alongside their keyboard. As a consequence, as they look down, they have to bend the head and neck forward. They will also have to rotate the head repetitively as they look from the document to the screen. Therefore, working in this manner is likely to result in discomfort in the neck. Research (Goostrey et al. 2014) has shown that placing source documents on the desk surface to the side of the keyboard produces both the greatest neck movement and muscle activity in all muscle groups compared to other options.

It is also the case that documents located on the desk surface are presented at a different viewing distance from that of the screen. Therefore, each time the user looks from the document to the screen and back again, they have to refocus their eyes. Such refocussing increases the workload for the eyes and many users might experience eye fatigue or headaches as a result.

To minimise the strain on the cervical structures, document holders have been designed to maintain the head and neck in a neutral posture, which is considered to be optimal to minimise neck extensor load (Chaffin and Andersson 2006). Pheasant and Haslegrave (2006) suggest that the postural loading on the neck muscles will be reduced considerably with the use of document stands. It would therefore appear to be sensible for document holders, or copy stands, to be provided in situations where an individual has to refer to documents while interacting with the screen or keyboard as a normal and consistent feature of their work.

Some may argue that the documents they work with are far too big to place on a document holder. Copy stands have been designed to accommodate a whole range of document sizes which should assist most users. In particular, document plinths that are intended to fit in the gap between the screen and the keyboard, presenting the documents in a tilted midway position, have been found to be extremely effective.

2.5.4 Screen Risers

The importance of presenting the screen at the appropriate height relative to the user was outlined in Chapter 1. However, many users find that they are offered only two choices when it comes to setting up their screen height: they can either have it on the desk, or they can have it sitting on the processor on the desk. For many female users, having the screen on the processor presents the screen at too high a level which causes them to work with their heads raised. For many male users, having the screen on the desk surface presents it at too low a level causing them to work with their heads and necks bent forward.

Screen risers can be purchased in a simple building-block platform style or in a more complex articulating arm form. Both styles can be equally effective. The building-block platform style allows the individual to add an additional level to the stack until the screen is at an appropriate height.

Articulating arms are usually adjustable for height, which enables users to set the height of the screen to suit their own preferences; the arms also usually allow users to pull the screen forward or push it back and sideways, which allows them to free up desk space when not performing screen-based work.

2.5.5 Telephone Headsets

Despite the fact that increasing use of email has reduced the need to make telephone calls, there are still many workers who have to talk on the telephone for lengthy periods or repeatedly through a working day. Although organisations that operate as dedicated call centres are set up to accommodate this work and distribute headsets to avoid the need to use telephone handsets, there are many offices where the handset is clamped between the ear and shoulder, usually as the individual inputs information into the system or accesses information on the screen. This results in static muscle tension and poor head position, which can lead to the potential for fatigue and discomfort in the neck (Pheasant and Haslegrave 2006). Telephone headsets or ear pieces should be made available to anyone who performs a task that demands using the telephone at the same time as interacting with the computer.

2.5.6 Reading Slopes

Some individuals will be involved in tasks that require them to read large quantities of documentation for extended periods. Many of these individuals spend long periods leaning forward over their desks with their heads lowered toward the source documents. This can, over time, lead to discomfort in both the neck and lower back. This situation can be resolved with the provision of a reading slope—sometimes called a reader's or editor's portable desk—assuming that the desk height cannot be altered easily. This slope simply raises the documents towards the user at a convenient angle (some are adjustable to several different angles), thereby reducing the need to lean forward.

Similar surfaces have been advertised by office equipment suppliers for use with keyboards or laptops. Care should be taken when considering the use of such equipment. The user should not be required to operate the keyboard either with the arms raised or the hands bent up at the wrist.

2.6 Summary

- Despite the fact that many brochures try to assure purchasers that their workstation furniture will accommodate many different users, as well as meeting any legal obligations, many desks fail from a usability/functionality perspective once introduced into a real working environment.

- Fixed-height desks, if used in conjunction with fully adjustable chairs, are perfectly acceptable for use in an office.
- The only individuals who may have difficulty in using fixed-height desks are very tall or short people and people with specific disabilities. As long as short people are provided with suitable adjustable chairs, they, too, should be able to use standard fixed-height desks.
- Tall individuals need adjustments to their desks in the form of either retrofit attachments to increase the leg length of the desk, or an alternative desk design, possibly a height-adjustable desk.
- Height-adjustable desks come in various forms. They can have telescopic legs that can be extended; they can have a crank handle that is rotated to raise or lower the desk; or they can be electrically-powered and adjusted using a small panel embedded in the desk surface.
- Features on the undersurface of the desk should not compromise an individual's sitting position.
- One of most important features of the desk is the surface area available. It should be sufficient to allow users to arrange their screens and keyboards one in front of the other and directly in front of them.
- The depth of the desk should be sufficient to allow users to position the screen about fingertip distance away from them when in their at-work position.
- The screen should not be positioned to one side of the seated operator so that it causes them to rotate their head to look at it or to twist their upper body. The user should always be able to sit directly in front of the screen and keyboard when they are in use.
- The styling of work surfaces should not compromise the way in which an individual works. Care should be taken that the work surface styling allows the equipment to be laid out correctly and should permit the user to adopt a comfortable working posture.
- The desk surface needs to be big enough to accommodate the mouse, documents, copy stand, and telephone, if in use.
- Desks should undergo properly managed trials before decisions are made about their purchase and introduction to the workplace to ensure that they suit the individuals and the task being performed.
- Leading edges on desks should be rounded or bull-nosed to prevent compression of tissue in the wrist area.
- The layout of the work surface should be such that all items used for extended periods or regularly fall within the normal work area are within easy reach. Items used on an irregular or infrequent basis can be stored in the zone of convenient reach, which can be reached with a fully extended arm.

- Items should not be stored on the floor around the seated user as this will require stooping or twisting to reach towards them. Such users should be provided with either a larger surface area or a trolley to accommodate the overflow.
- The undersurface of the desk, both front to back and side to side, should allow users to move their legs freely and change position without hindrance.
- The location of drawers should be carefully considered so that they do not limit the control users have over where they sit at a desk.
- Cable management should be easy to use so that it does not discourage people from storing cables in the management system after desks have been moved around the office or equipment has been repositioned. Cables should not lie around the user's feet.
- Cable should be of a sufficient length to allow for standing desks to be raised.
- Full consideration of task demands should be made before partitioning is introduced, as well as the likely impact on the psycho-social element.
- Partitions can result in isolation and can have a negative impact on group cohesion.
- Partitioning is most suited to jobs requiring confidentiality or to individuals who use the phone frequently, which may distract other workers.
- Computer users should not be required to sit on anything other than a fully adjustable chair when working.
- The chair should be adjustable for height.
- The backrest should be adjustable for tilt.
- The lumbar support on the backrest should be capable of being repositioned relative to the lumbar area of the seated user.
- The chair should have a five-star base on casters.
- The chair should have sufficient padding to ensure the user is comfortable.
- High backrests and head supports on chairs are not an absolute necessity, and careful consideration should be given to whether they are required before they are introduced.
- Head supports are usable only when the head is rested on their surface, which makes reference to the screen more difficult.
- Care should be taken that smaller users have not been provided with chairs so large that they find it difficult to sit back against the backrest.

- Armrests on a chair should not cause an obstruction when the user tries to sit close to the leading edge of the desk.
- Users should not lower their chairs in order to get their armrests under the desk so they can sit closer to the work surface. This will cause them to sit too low relative to the desk and keyboard.
- Problematic armrests should be removed.
- Ideally, armrests should have a degree of adjustability both for height and width settings.
- Any adjustment mechanism on a chair should be easy to use from a seated position. Any adjustment that requires the user to stand up or is complex will not be used properly, making it unlikely that the user will adopt an appropriate sitting position.
- Users should be given thorough training in the use of their chairs, the adoption of appropriate working postures and how to use their workstations.
- Footrests should be given only to users whose feet do not touch the floor once they have adjusted the height of their chairs.
- A footrest needs to have a large surface area; it needs to be covered in non-slip material; and, ideally, it should be height- and tilt-adjustable.
- Users should not lower their chairs to get their feet on the floor. This causes the height of the keyboard and desk to rise relative to their seated position and forces them to work with their arms raised.
- If wrist rests are provided for use in front of the keyboard, users should be advised not to lean on these while they are keying. This will cause them to overextend their fingers.
- Copy stands or document plinths should be provided for individuals who work from documents while interacting with the computer.
- The screen should be set at a height suitable for each individual who might sit at the desk. The most effective way of doing this is by using either an articulating arm or building-block style platforms.
- Individuals who are required to make lengthy or regular phone calls should be provided with headsets. They should not clamp the handset of a telephone between shoulder and ear.
- Individuals involved in lengthy periods of reading or writing should be provided with a slope that raises the documents up towards them. This will eliminate the need to bend forward.

References

Chaffin, D. B. and Andersson, G. B. 2006. *Occupational Biomechanics*. 4th ed. New York: Wiley.

Chau, J., van der Ploeg, H., van Uffelen, J., Wong, J., and Riphagen, I. 2010. Are workplace interventions to reduce sitting effective? A systematic review. *Preventive Medicine* 51 (5), 352–356.

De Croon, E. M., Sluiter, J. K., Kuijer, P. P. F. M., and Frings-Dresen, M. H. W. 2005. The effect of office concepts on worker health and performance: A systematic review of the literature. *Ergonomics* 48 (2), 119–134.

Delleman, N. J. 2004. Head and neck. In N. J. Delleman et al. (Eds.), *Working Postures and Movements: Tools for Evaluation and Engineering*. Boca Raton, FL: CRC Press, pp. 87–108.

Goossens, R., Netten, M., and van der Doelen, L. 2012. An office chair to influence the sitting behavior of office workers. *Work* 41 (1), 2086–2088.

Goostrey, S., Treleaven, J., and Johnston, V. 2014. Evaluation of document location during computer use in terms of neck muscle activity and neck movement. *Applied Ergonomics* 45, 767–772.

Healy, G. N., Eakin, E., LaMontagne, A., Owen, N., Winkler, E. A. H., Wiesner, G., and Dunstan, D. 2013. Reducing sitting time in office workers: Short-term efficacy of a multicomponent intervention. *Preventive Medicine* 57 (1), 43–48.

Karakolis, T., Barrett, J., and Callaghan, J. P. 2016. A comparison of trunk biomechanics, musculoskeletal discomfort and productivity during simulated sit-stand office work. *Ergonomics* 59 (10), 1275–1287.

Karakolis, T. and Callaghan, J. P. 2014. The impact of sit-stand office workstations on worker discomfort and productivity: A review. *Applied Ergonomics* 45 (3), 799–806.

Lin, M. Y., Barbir, A., and Dennerlein, J. T. 2017. Evaluating biomechanics of user-selected sitting and standing computer workstation. *Applied Ergonomics* 65, 1–7.

Lin, M., Catalano, P., and Dennerlein, J. 2016. A psychophysical protocol to develop ergonomic recommendations for sitting and standing workstations. *Human Factors* 58 (4), 574–585.

Marshall, P. W., Patel, H., and Callaghan, J. P. 2011. Gluteus medius strength, endurance, and co-activation in the development of low back pain during prolonged standing. *Human Movement Science* 30 (1), 63–73.

McKeown, C. 2011. *Ergonomics in Action: A Practical Guide for the Workplace*. Boca Raton, FL: CRC Press.

Nelson-Wong, E. and Callaghan, J. P. 2010. Is muscle co-activation a predisposing factor for low back pain development during standing? A multifactorial approach for early identification of at-risk individuals. *Journal of Electromyography and Kinesiology* 20 (2), 256–263.

Netten, M., van der Doelen, L., and Goossens, R. 2011. Chair based measurements of sitting behavior: A field study of sitting postures and sitting time in office work. In *Digital Human Modeling and Applications in Health, Safety, Ergonomics, and Risk Management. Human Body Modeling and Ergonomics*. Berlin, Germany: Springer, pp. 261–268.

Pheasant, S. and Haslegrave, C. 2006. *Bodyspace: Anthropometry, Ergonomics and the Design of Work*. 3rd ed. Boca Raton, FL: CRC Press.

Pronk, N. P., Katz, A. S., Lowry, M., and Payfer, J. R. 2011. Reducing occupational sitting time and improving worker health: The take-a-stand project. *Preventing Chronic Disease* 2012 (9), 110323.

Robertson, M. M., Ciriello, V. M., and Garabet, A. M. 2013. Office ergonomics training and a sit-stand workstation: Effects on musculoskeletal and visual symptoms and performance of office workers. *Applied Ergonomics* 44 (1), 73–85.

Ryan, C., Grant, P., and Granat, M. 2011. Sitting patterns at work: Objective measurement of adherence to current recommendations. *Ergonomics* 54 (6), 531–538.

Sengupta, A. K. and Das, B. 2000. Maximum reach envelope for the seated and standing male and female for industrial workstation design. *Ergonomics* 43 (9), 1390–1404.

Szeto, G. P. Y., Straker, L. M., and O'Sullivan, P. 2005. A comparison of symptomatic and asymptomatic office workers performing monotonous keyboard workd1: Neck and shoulder muscle recruitment patterns. *Manual Therapy* 10, 270–280.

Thorn, S., Sogaard, K., Kallenberg, L. A. C., Sandsjo, L., Sjogaard, G., Hermens, H. J., Kadefors, R., and Forsman, M. 2007. Trapezius muscle rest time during standardized computer work e a comparison of female computer users with and without self-reported neck/shoulder complaints. *Journal of Electromyography and Kinesiology* 17, 420–427.

Van Dieën, J. H., De Looze, M. P., and Hermans, V. 2001. Effects of dynamic office chairs on trunk kinematics, trunk extensor EMG and spinal shrinkage. *Ergonomics* 44 (7), 739–750.

Wilks, S., Mortimer, M., and Nylen, P. 2006. The introduction of sit-stand worktables; aspects of attitudes, compliance and satisfaction. *Applied Ergonomics* 37 (3), 359–365.

Wittig, T. 2000. Ergonomische Untersuchung alternativer Büro- und Bildschirmarbeitsplatzkonzepte. 1. Auflage. Wirtschaftsverlag NW: Verlag für neue Wissenschaft GmbH, Bremerhaven. (Schriftenreihe der Bundesanstalt für Arbeitsschutz und Arbeitsmedizin: Forschungsbericht, Fb 878).

3

Workstation Trials

3.1 Introduction

Many an organisation has wasted significant money by investing in inappropriate or poorly designed workstation furniture. Glossy brochures sell the idea that the furniture will meet every possible need that an employer might have. Office furniture showrooms present a pristine environment filled with shiny new desks and attractive chairs. In both cases, the superficial qualities of the equipment can have an immediate impact on decision-making. Unfortunately, decisions made on the basis of a few minutes of viewing or testing will not provide a clear indication of how successful that equipment might be once used in a real work environment. Getting a true appreciation of the likely success of any piece of equipment—whether a desk, chair, footrest, document holder, or anything else—is possible only after the equipment has been used during the performance of normal work in the real environment. This illustrates the importance of having equipment on trial before committing to purchasing it, particularly if large numbers are involved.

If trials are to yield helpful information, they need to be designed and managed carefully so that usable feedback can be collected. The feedback is the means by which an organisation will identify accurately whether selected equipment is suitable for the work being performed and whether the workers actually like it.

Trials can last a number of hours, several days, or even weeks. The duration and complexity of the trial is very much reliant on the nature of the equipment on trial, its likely impact once introduced into the working environment, and the number of individuals who might use the equipment. For instance, a reading slope intended to be employed by one person might be used on a trial basis by the intended user for just a few hours. In contrast, an organisation undertaking refurbishment may intend to introduce new desks and chairs for 800 employees, which would require a much more involved process of selection, short-listing, and trialling.

Before starting the process of selecting furniture to include in a trial, the individual given responsibility for drawing up the shortlist should be in a position that ensures a complete understanding of what the users' tasks entail and what those users need to complete their tasks accurately, efficiently, safely, and comfortably. The trial manager should then prepare a list of requirements to offer potential equipment suppliers.

Some of the methods for generating an understanding of what the users' work involves include direct discussions with the individuals performing the work and their managers or team leaders, reading job descriptions or standard operating procedures, conducting surveys, and conducting a task analysis.

3.2 Task Analysis

A task analysis is a formal method for both describing and analyzing the performance demands placed on an individual within the working environment. It is distinct from a job analysis, which provides a breakdown of the activities, on a more general level, performed by a particular individual within an organisation. A job analysis usually takes the form of a list of activities, duties, or responsibilities associated with the particular role under scrutiny. As part of the process of task analysis, complete tasks are systematically broken down into components and subcomponents; the result is referred to as the 'task description'. This can be achieved successfully only by observing a skilled individual over an extended period, while recording each activity or movement.

At the end of the analysis, a profile will have been generated that specifies: (1) the task requirements, such as operating the keyboard (possibly differentiating between alpha and numeric work), using the mouse, answering the telephone, reading documents, using a pen, and so on; (2) the task environment, such as operating the mouse on the desk surface; and (3) the task behaviour, which provides insight into how the user is working—for example, reaching across the desk to the mouse or holding a pen in the hand while inputting numerical data. Task behaviour is influenced by previous experiences and training and limited by psychological and physiological factors.

In addition to specifying exactly what the individual's tasks involve, the task analysis provides a structured breakdown of how long is spent on each component or subcomponent and what percentage of the overall task that aspect represents. One goal of this process is to enable the employer to identify the priorities in the task by recording either what element of the task is performed for the longest period of time or which element of the task is performed most often throughout the working period. By carrying out this procedure, the employer will get a profile of what any new workstation furniture has to achieve in terms of enabling users to complete their work effectively and efficiently. Also, developing a methodical sequence of task demands enables employers to compare the sequence with what is already known about human capabilities. This process allows them to identify inefficiencies in the current work—inefficiencies, in this context, meaning ergonomic inefficiencies where the user is working in a manner that is not ideal. For example, the task analysis may highlight that a significant feature of the work is telephone answering. Reference to the desk layout may reveal that the telephone is located at the most distant point on the desk, which results in repeated reaching as the user picks up the handset. In addition to highlighting the poor workstation layout, the task analysis might pinpoint the need for a headset or ear piece as an alternative to holding the handset of the telephone to the ear repeatedly for extended periods.

The task analysis can also aid in identifying the skills required to complete the tasks successfully or, perhaps, the training needed to develop the skills. Obviously, to provide such information, the task analysis should be thorough and include all components and subcomponents of the tasks. This underlines the importance of taking the time to generate a thorough task description prior to starting the analysis. The description details the step-by-step actions and movements involved in the operation being observed.

3.2.1 Starting the Procedure

At the outset of the task analysis process, there should be agreement on what the analysis should achieve. This will allow decisions to be made about how the analysis will be carried out. For instance, how many individuals will be observed, and how many, at what level, within the organisation? To establish exactly who would be considered representative of the organisation's working population, the appropriate personnel with access to this information should be involved in the decision-making and planning process from an early stage. The users who will be the focus of the observations need to be informed about the process, and confirmation needs to be given that each relevant individual will be available in the specific period when the task analysis is to be carried out. Consideration has to be given to the fact that even individuals identified as performing identical tasks may actually perform them differently because of personal preferences, and this may have an impact on the results of the analysis.

Having established the aims of the task analysis at the outset of the process, those compiling the data will be able to collect information that is directly relevant to the project's aims. To maintain this, they also need to ensure that they detail the task components that make up the task description accurately. This suggests that cross-referencing with other individuals performing the same task is advisable.

Consideration should be given to whether more than one type of operation is performed at the workstation at different times. For instance, an individual might perform dedicated screen-based operations on some days, but clerical duties on other days. Attention needs to be paid to whether more than one operator will use the same workstation during a 24-hour cycle, as with a multiple shift system in a call centre. Steps need to be taken to ensure that a representative sample is covered during the analysis. If particular tasks cannot be observed for practical reasons, efforts should be made to ensure that these do not contain elements likely to undermine the findings of the actual observations. This can be achieved through discussions with relevant personnel and with reference to any documentation that records details relating to the tasks.

Decisions have to be made prior to commencing the analysis about whether the individual carrying out the observations will interrupt the worker being observed to clarify points or whether they will wait until the end of the observations and then carry out a 'debriefing' session. In the former case, the operation can be performed and observed in real time, which offers a more realistic view of the demands placed on the individual. In the latter case, debriefing should be carried out soon after the observations have ceased to avoid any effects resulting from delayed information recall by the worker.

If task descriptions—the list of individual components making up the overall task's demands—are being developed on the basis of documentation such as job descriptions or training materials, it should be kept in mind that these documents may not record every step in a sequence of steps; they may not provide sufficient detail to allow for a proper analysis; the contents may include ambiguous wording; or the writer may have made a number of erroneous assumptions about the operation. In addition, the contents of the documents may be out of date even though the document itself has been produced quite recently, perhaps because the document was based on information that was collated some time before the document's creation.

Task descriptions can be generated through discussions with personnel, either through formal or informal conversations with individuals or by administering a questionnaire. Care should be taken to select the personnel who are likely to offer the most accurate information relating to the task demands. That is not to say that only experienced employees should be involved in this process. Long-term employees may have developed working practices that would not be deemed appropriate and would not be advisable to perpetuate, even though they themselves may perceive their own working practices as 'normal' and acceptable. Clearly, designing and selecting a workstation to accommodate poor working practices would not be an advisable strategy.

3.2.2 Information Collection

The driving force behind the observations should be that all information collected will address the issues of concern. Those collecting the information should not be open to criticism that they were biased in the sense that they focussed only on task features that they considered important. A record should be made of exactly what has been observed, in what environment, with what equipment, and who was performing the task. Creating formal records contributes to the development of an audit trail that later could become important in illustrating how purchasing decisions were made and the rationale behind them.

During the course of the data collection, if the analyst identifies any conflicts in terms of how they understand the task should be completed, having read documented explanations and what they are observing, they should enquire why such conflicts are apparent. If the session is being timed, the questions should wait until the session is over so that the user is not interrupted. To avoid any confusion or conflicts during the actual data collection, it is always useful for the analyst to observe an untimed run-through of the task during which queries can be addressed.

If an individual is describing a task during its execution as a means to expand on the data collected by the analyst, the worker may need to be encouraged to offer sufficient detail about actions and movements. To facilitate this, the worker should be given an explanation of the purpose behind the task analysis. The analyst needs to be alert to the fact that the worker may, during the description of the task, move away from the focus of the analysis and start to discuss unrelated or irrelevant topics. Alternatively, if the analyst becomes aware that the user is reluctant to describe the tasks or becomes hesitant during the discussions, the analyst should investigate this. Group discussions might give individual users more confidence when providing information. However, care needs to be taken that senior members in the group do not monopolise the discussions.

When the actual observations are being carried out, it is quite useful to make a video recording of the operations, as well as observing directly. This allows clarification of details at a later point and avoids the need to request the user to repeat an operation or movement. Many people, however, become self-conscious when being videoed, and this may have an effect on performance. For that reason, it is advisable to video for a longer period of time than might be considered necessary. This will allow the person being videoed sufficient time to settle into working and forget about the camera.

3.2.3 Recording Information

Having acquired a full understanding of the task, the analyst will record its constituent parts, breaking it down continually into a series of subtasks until they believe they have detailed every element. This is referred to as a 'hierarchical task analysis' (HTA). Once the full list has been generated, the analyst can then augment this with additional information, such as the time taken to execute each element of the task. Table 3.1 shows a simple example of a task analysis sheet completed during the observation of a secretarial task. The sheet also records the frequency with which individual subtasks are performed and the length of time spent on any one subtask throughout the period of observation.

TABLE 3.1
A Simple Example of a Task Analysis Sheet Completed During the Observation of a Secretarial Task

TASK ANALYSIS SHEET

User Name: J. Pelling **Job Title: Secretary** **Date: 14th September**

Task Activity:	Task commenced at:																Total time spent on activity:	Percentage of overall cycle:	Frequency:
Keyboard use	0		1.05		3.27				7.35		11.02						13 mins 23 secs	90.7	4
Mouse use		0.55								10.47		14.31					33 secs	3.7	3
Telephone use				3.20													7 secs	0.8	1
Reading documents						6.59		7.21									29 secs	3.3	2
Write on documents							7.14										7 secs	0.8	1
Reach to printer													14.39				6 secs	0.7	1
Leave desk														14.45					1

Acquiring information about what is being done and how it is being done will help to highlight any mismatches between what the individual is required to do and what facilities they have been given to complete their work. This will enable the development of a clear understanding of needs for future equipment provision so that the equipment is selected, specifically, to meet the needs of the user.

3.3 Trials

The task analysis process enables individuals responsible for producing specifications for office equipment to generate an accurate list of requirements that will adequately accommodate the user's needs. For instance, they might suggest a desk unit with a particular surface area, or a keyboard without a number pad, if the task analysis highlighted that the integral number pad was not being used. This list can be given to the equipment suppliers who should then be in a position to state whether they have any equipment that matches those requirements. The suppliers are likely to present an inventory of possible solutions to the decision maker.

Once suppliers have provided the inventory to the decision maker, a shortlist needs to be drawn up. It is always preferable to assess all suggested items of furniture in person, rather than by simply reading a spec, before drawing up a shortlist. This will help to weed out desks or chairs with design inadequacies that are not obvious from advertising material, or have not been mentioned (or even recognised) by the supplier. Although this may sound like an enormous task, the process of elimination is not always hugely time-consuming, since some equipment will be obviously unsuitable. In addition, when confronted with a series of chairs by one supplier, it is worth bearing in mind that a comparison between seats, ranking them in order of comfort to discomfort, can actually be made within a few minutes relatively successfully. This might provide an initial starting point for creating a shortlist.

Once the shortlist has been generated, arrangements should be made for all of the items to be delivered to the employer's site so that they are all available for inclusion in a trial. The trial allows testing of the equipment under controlled conditions. These conditions should be as realistic as possible, because it is the combination of the user performing tasks in conjunction with the equipment that creates the 'interface', and it is the interface that is being evaluated. The trial allows the interaction between the equipment and the user to be systematically observed and measured. The use of the term 'trial' suggests that the equipment design is in its final form and will be introduced into the workplace without change once the exercise is complete, assuming it is identified as suitable. 'Prototyping', on the other hand, is a phase employed to describe testing a suggested design prior to the final trial.

This is likely to be done in situations where a workstation design has been tailored specifically to meet the needs of a particular individual or operation within a workplace. For instance, in a control room environment that employs closed circuit television (CCTV), a standard desk is unlikely to be appropriate. Therefore, a custom-made workstation design might be required. Before this is engineered and constructed in its final form, a prototype can be tried out. This process involves the creation of the unique workstation in rough form, for instance, in chipboard. This enables the user to test the dimensions and layout of the suggested design before making the commitment to invest in the full cost of producing the workstation in the final form. Modifications can be made to the design to make it more suitable for the intended task without incurring the expense of commissioning a whole new finalised workstation.

Individuals selected for participation in the trial should be representative of those who are likely to use the equipment in the future in terms of the type of work they perform. They also need to have characteristics, both physical and psychological, that are considered representative of the likely user population. The types of characteristics to consider include general characteristics, such as age and gender; physical characteristics, such as height, weight, motor control, and visual and auditory capabilities; cognitive characteristics, such as problem-solving and decision-making; personal characteristics, such as motivation and attitude; and experience characteristics, such as training, skill base, and familiarity with equipment in use. To achieve this successfully, a profile of the type of people who will be found in the user population should be developed. Emphasis should be placed on the elements of the profile that are relevant to the trial itself. Where possible, how the people in the user group vary across a particular feature should be indicated—for instance, the range of heights among users.

The number of users included in a trial within a working environment is usually driven by the number of individuals who are prepared to participate in it and the circumstances prevailing at the time the trial is scheduled to take place. Circumstances may limit the number of individuals who are free to participate; however, to achieve valid results, a sufficient number needs to be included to allow for accurate statistical analysis. If an organisation has no choice but to have only a small number of individuals in the trial, it should be expected that even a small number of individuals is likely to be sufficient to identify the most extreme problems with new equipment.

Having selected the individuals who will participate, trial managers must consider the type of work that should be performed during the trial. The aim should be to use the equipment under conditions that are as nearly normal as possible. However, many users will perform a variety of tasks at different points in the day or at different times of the week. The trial manager should ensure that whatever work is performed during the trial, all of the attributes of the proposed workstation equipment should be used so that the equipment is properly tested and that the product use is typical of what would occur when not under trial conditions. For instance, the sequence of tasks performed should be logical and typical of what would be performed in the non-trial setting.

The trial needs to be scheduled with a strict timetable so that all those participating know exactly when they will be using any particular piece of equipment, and for how long. The procedure followed should be consistent for each individual participating in the trial.

To run trials effectively, it is advisable to minimise disruption of the employees' work. To that end, if each employee participating in the trial is required to try several chair and desk combinations, it is helpful to schedule the trial of any combination so that it coincides with the first part of the working day or the second part of the working day. In effect, each combination would be used for about three or four hours, as this is likely to produce a stable rating of the comfort of a work seat.

Prior to commencing the trial, all participants need to be thoroughly briefed so that they understand exactly what they were committing to, how the trial is to be run, and how their feedback is going to be recorded. This has to be backed up by a trial manager, who will be available during the trial, to remind individuals about the process involved. The trial manager should be a good communicator and be able to build a good rapport with the users quickly.

Information from the trial can be collected during interviews with the users, through questionnaires, by direct observation of the users when working with the equipment on trial, or by the collection of objective data. Information collection does not have to be a complicated, involved process. Useful information can be collected using very simple methods, such as basic rating scales.

Traditionally, interviews are carried out on a one-to-one basis; however, group interviews are more common when products are being evaluated. At the start of the interview, users should be put at ease before the critical data collection questions are posed. They should then be steered subtly so that they offer a general evaluation of the product and describe any difficulties they have encountered when using it. Users tend to remember and focus on negative experiences more than positive ones, so it is important to encourage them to relay positive experiences as well as negative ones. They may need to be prompted to evaluate attributes of the equipment that, under normal circumstances, they would not normally think about. The trial manager needs to be able to make the users aware of the product attributes without influencing the users' impressions of them. This process can be assisted by carrying the interviews out in close proximity to the equipment used during the trial so that users have an opportunity to consider particular characteristics previously overlooked, or so that they can demonstrate important points. During the interview, users should also be encouraged to suggest how the product might be improved. They also need to feel that their contribution is worthwhile.

Questionnaires can be used to collect information relating to the use of the equipment and the users' perception of that equipment. Questionnaires, or feedback forms, can include open-ended or closed questions. Open-ended questions do not limit the answer that can be given by the user. For instance, the questionnaire may ask users what they thought about the ease with which a chair could be adjusted. Answers to such an open-ended question require more involved and time-consuming post-trial analysis and coding by the trial manager. A closed question limits the choices that users have with regard to the answer, and this is the process employed in many types of surveys. Users are provided with a predetermined set of answers to each question, and they select one answer that best describes their views on a particular attribute of the equipment. Closed questions may simply require a 'yes' or 'no' answer. Rating scales can be used to measure the user's views on equipment characteristics, or a ranking scale can be used to allow the user to place equipment in order of preference with regard to particular attributes.

Table 3.2 provides an example of a feedback form that can be used during a trial involving new desks. Table 3.3 gives an example of a feedback form that can be used during an extended trial involving individual seats. Table 3.4 shows a feedback form that can be used during a simultaneous comparison of three seats.

The location where the trial occurs will be determined by a number of variables, such as space. Wherever it is located, the trial manager should ensure that the context in which the trial takes place is representative of the context in which the user works. For example, the environmental features of an office, such as lighting levels, the temperature, humidity, and noise level all influence the user's overall perception of their working environment. For that reason, it is always preferable to conduct the trial in the user's own working environment.

One of the major failings of trials can result when a full evaluation of the collected information does not take place. In that case, an opportunity to identify faults in the design will be lost. A thorough and systematic analysis of the data collected will provide the trial manager with a strong indication of the likely suitability of any equipment included in the trial.

TABLE 3.2

Sample Feedback Form for Use during a Trial of New Desks

DESKING TRIAL

It is **ESSENTIAL** you read the information below
BEFORE you answer the questions overleaf

Purpose of the trial

It is a very important part of the furniture selection process to seek the opinions of those who will use it. This trial is just one of several stages of seeking such opinions.

There are a number of different workstations on trial and we would like you to evaluate just three, selected at random, and use each of them for half a day. In your case we want you to compare workstations:

What we want you to do

These trials are to compare 11 key usability features of the workstations.

To make the comparison fair it is important you adopt a standard posture at each workstation – as illustrated below. To do this:

1. Sit in front of your computer as normal and arrange the keyboard and mouse so they are close to you.

2. Set your chair height so that your elbow is level with the middle row of the keyboard and your feet are flat on the floor or footrest.

3. Adjust your chair so you are fully supported by the backrest.

From this 'standard' starting posture, please compare the 11 features opposite and identify how the workstation in use scores on a scale of one to seven as follows.

1 : Very good
2 : Good
3 : Slightly good
4 : Satisfactory
5 : Slightly poor
6 : Poor
7 : Very poor

Complete one section of this form after you have used the first workstations for half a day. Repeat after using the second and third workstations for half a day.

(Continued)

TABLE 3.2 (*Continued*)

Sample Feedback Form for Use during a Trial of New Desks

	X	Y	Z
Space for keyboard *Ability to arrange the keyboard in a comfortable position relative to you, your mouse and your monitor.*			
Space for mouse *Ability to position the mouse for comfortable and unhindered use.*			
Space for monitor(s) *Ability to position the monitor(s) at a comfortable viewing distance, position and angle to minimise head and neck movements.*			
Space on the workstation *Space to arrange desktop items such as telephones to minimise frequent head and neck movements, overstretching or without feeling cramped.*			
Space for paperwork *Ability to position papers, reference material etc. in comfortable and convenient positions.*			
Thigh clearance beneath worksurface *Ability to sit at all areas of the workstation without catching your thighs on the underside.*			
Knee clearance beneath worksurface *Ability to sit at all areas of the workstation without your knees coming into contact with anything on the underside or underneath.*			
Lower Leg clearance beneath worksurface *Ability to sit at all areas of the workstation with your lower legs in a comfortable and uncramped position.*			
Access to all areas of the workstation *Ability to move freely across all areas of the workstation (e.g. to access your pedestal) without being impeded by anything on the underside or underneath.*			
Appearance of the workstation *Consider the style, shape and design of the worksurface, legs and frame. Do not consider the materials and colour - they will be selected later.*			
Overall suitability *All things considered, compare each workstation for suitability for the work you do.*			

General comments

TABLE 3.3

Sample Feedback Form for Use during an Extended Trial Involving Individual Seats

SEATING TRIAL

It is **ESSENTIAL** you read the information below **BEFORE** you answer the questions opposite

Purpose of the trial

It is a very important part of the seating selection process to seek the opinions of those who will use it. This trial is just one of several stages of seeking such opinions.

There are too many different seats on trial to expect everyone to evaluate them all. Therefore, we are asking individuals to compare just three, selected at random. In your case we want you to compare seats:

What we want you to do

These trials are to compare key features of the seats. To make the comparison fair it is important you adopt a 'standard' posture for each seat - as illustrated below. To do this:

	1. Sit in front of your computer as normal and arrange the keyboard and mouse so they are close to you. 2. Set your chair height so that your elbow is level with the middle row of the keyboard and your feet are flat on the floor or footrest. 3. Adjust your chair so you are fully supported by the backrest.

Complete one questionnaire after you have used the first chair for half a day.

Complete further questionnaires after using the second and third chairs for half a day.

Allow approximately 15 minutes to complete the questionnaire.

(Continued)

TABLE 3.3 (*Continued*)

Sample Feedback Form for Use during an Extended Trial Involving Individual Seats

All responses will be treated as confidential. We do need to know your name and job title just in case we need to contact you about any of your answers.

ABOUT YOU

Name:	**Job Title:**

Contact telephone number:	

ABOUT THE CHAIR

Are the seat adjustment controls easy to reach and operate from your seated position?		
Seat Height Adjustment	Yes ☐	No ☐
Back Height Adjustment	Yes ☐	No ☐
Back Angle Adjustment	Yes ☐	No ☐
Seat length Adjustment	Yes ☐	No ☐
Tension Adjustment	Yes ☐	No ☐
If **'No'**, please give details: ________________		

Is the range of adjustment provided adequate?		
Seat Height Adjustment	Yes ☐	No ☐
Back Height Adjustment	Yes ☐	No ☐
Back Angle Adjustment	Yes ☐	No ☐
Seat length Adjustment	Yes ☐	No ☐
Tension Adjustment	Yes ☐	No ☐
If **'No'**, please give details: ________________		

If your chair has armrests are they comfortable?	Yes ☐	No ☐
Is the seat stable?	Yes ☐	No ☐
Do the castors operate well?	Yes ☐	No ☐
Is the seat aesthetically pleasing?	Yes ☐	No ☐
Does the seat feel and look well designed?	Yes ☐	No ☐

(*Continued*)

TABLE 3.3 (*Continued*)

Sample Feedback Form for Use during an Extended Trial Involving Individual Seats

Please mark on the illustration below, any areas on your seat that you have found to be 'TOO HARD' or 'TOO SOFT' for comfortable sitting.

Does the seat provide what you would consider to be good support?	Yes ☐ No ☐ **(If 'No' Please give details below)**

Please mark on the figure below any body areas where you have felt discomfort whilst using the chair.

Please detail the type of discomfort you experienced (e.g. pins and needles, numbness, sweating, aching, pain, etc.) and if possible the cause of the discomfort.

Please use the space below for further comments you would like to make about your chair.

TABLE 3.4
Sample Feedback Form for Use during the Simultaneous Comparison of Three Seats

SEATING TRIAL

It is **ESSENTIAL** that you read the information below
BEFORE you answer the questions overleaf

Purpose of the trial

It is a very important part of the seating selection process to seek the opinions of those who will be using them. This trial is just one of several stages of seeking such opinions.

There are too many different seats on trial to expect you to evaluate them all. Therefore, we are asking you to compare just three, selected at random. In your case we want you to compare seats:

D E & F

What we want you to do

These trials are to compare key features of the seats. You will be shown how to adjust each seat prior to you completing the evaluation. Once you are familiar with the adjustment mechanisms we would like you compare the three seats with each other and rank them in order of preference.

For each seat you will be asked to indicate which one you thought was 'best', 'middle' or 'worst'.

We realise that these decisions may be difficult to make, but please answer all questions and do not rank any of the seats equally for any feature. Simply enter the identification letter, which can be found on the back of the seat, under the appropriate column to indicate your preferences.

Allow approximately 15 minutes to complete the questionnaire.

All responses will be treated as confidential. We do need to know your name and job title just in case we need to contact you about any of your answers.

About you

Name:	
Job Title:	
Contact telephone number:	
Approximate height:	
Approximate weight:	

(Continued)

TABLE 3.4 (*Continued*)

Sample Feedback Form for Use during the Simultaneous Comparison of Three Seats

About the seats

Are the seat controls easy to reach and operate from the seated position?	BEST	MIDDLE	WORST
Seat cushion height adjuster			
Seat cushion angle adjuster			
Seat cushion length adjuster			
Backrest height adjuster			
Backrest angle adjuster			
Backrest tension adjuster			
Armrest adjuster			
Any comments?			

Is the range of adjustment provided adequate?	BEST	MIDDLE	WORST
Seat height adjustment range			
Seat angle adjustment range			
Seat cushion length adjuster			
Backrest height adjustment range			
Backrest angle adjustment range			
Backrest tension adjustment range			
Armrest adjustment range			
Any comments?			

General issues:	BEST	MIDDLE	WORST
Does the seat move around easily on its castors?			
Is the seat stable?			
Are the armrests comfortable?			
Is the seat aesthetically pleasing?			
Does the seat feel and look well designed?			
Is the seat supportive and comfortable?			
Any comments?			

Overall	BEST
Of all the seats, which do you think is the most attractive?	
Any general comments on matters not covered elsewhere?	

Thank you for completing this questionnaire.

3.4 The Roll-Out

Having selected the equipment based on the results of the trial, the organisation has to consider how it will introduce the new product into the workplace. There are a number of strategies available that will determine, to an extent, the ease with which the workforce will accept the changes. It should be kept in mind that despite the fact that the new equipment is intended to improve the working environment, many people will be reluctant to accept the need for change, even if seemingly minor. Many of them will find the experience stressful, because they may perceive the changes to be out of their control; and this may create feelings of uncertainty.

There is a tendency for employees to accept a completely new environment more readily than changes to an existing environment. Unfortunately, few organisations have the luxury of starting with a 'greenfield' site where everything is new. Most organisations undergoing trials of new equipment are likely to be adding to what is already in place. Some of these organisations will wait until staff finish work on a Friday night, then have the refurbishment team work over the weekend changing all of the desks, chairs, screens, keyboards, and so on. When the staff come back to work on Monday morning, their environment has changed utterly. The mistake made by many of these organisations is to expect their staff to resume working as if nothing had happened. Even when the environment has been changed as a means of improvement, employees need to be given an acclimatisation period to get used to the changes. From the body's perspective, everything has changed: the postures adopted are likely to be different; the support offered by the chair is likely to be different; the feel of the keyboard will be different, and so on. Sudden changes, like these, without a reduction in the pace of work are often associated with the sudden onset of symptoms of upper limb disorders, typically, pain, tingling, and swelling in the arms and hands.

Some organisations adopt a more evolutionary process in which changes are introduced gradually, as opposed to the revolutionary approach. The only problem is that the evolutionary approach can result in the process of change dragging on and becoming fragmented.

Whatever approach is adopted, the implementation phase is likely to be most successful if the employees are involved in the process. Including them in equipment trials is one of the steps in the inclusive approach. There should also be briefings and discussion groups that include those not involved in the trials so that they understand what is occurring and why it is necessary. This also allows them the opportunity to ask questions.

If organisations do face resistance, this will be a result of individual factors, such as perceptions and past experience, organisational factors, such as lack of information, and the implications of the change, such as how the users feel they might lose out in the change process. People will weigh up the positive aspects and the negative aspects of change, and this evaluation will result in resistance, if the company does not take a proactive stance in combatting potential negative perceptions. Communicating with the workforce, increasing their understanding of the process, including them by involving them in trials, possibly offering incentives (not necessarily financial), and providing empathetic support are the means to overcome resistance.

3.5 Summary

- A true appreciation of the likely success of any new piece of equipment can only be gained if it is used for normal work in a real environment. Trials are a means of establishing whether workstation equipment is suitable.
- Trials should be designed and managed carefully, and usable feedback should be collected during the trials.
- Feedback is the means by which an organisation accurately identifies whether selected equipment is suitable for the work being performed and the workforce using it.
- The duration and complexity of the trial are determined by the nature of the equipment on trial and its likely impact on the working environment once introduced.
- Prior to creating a shortlist of possible furniture, an organisation needs to establish exactly what users' tasks entail and what they need in order to complete their work efficiently, safely, and comfortably.
- A task analysis is one of the most effective ways of generating an understanding of what the work involves.
- A task analysis is a formal method for describing and analyzing performance demands. During a task analysis, tasks are systematically broken down into components and subcomponents.
- The profile generated at the end of the task analysis identifies the task requirements, the task environment, and the task behaviour.
- Generally speaking, a task analysis provides an employer with a priority breakdown of the most important features of a job in terms of the frequency with which it is carried out or the length of time it is performed.
- The task analysis allows employers to identify inefficiencies in the current work and mismatches between task demands and workstation design and layout.
- Users involved in the task analysis process need to be representative of the organisation's working population.

- Several workers selected for observation during the task analysis may be performing the same work, but they may choose to complete it differently. For that reason crossreferencing with the manner in which other individuals work is advisable.
- Prior to the task analysis being carried out, a record should be made of exactly what has been observed in an environment, with what equipment, and exactly who was performing the work.
- Video recording of the observations during the task analysis is helpful because it allows clarification of points after the analysis has come to an end without the need to request the user to repeat the operations.
- Once the observations are complete, the analyst can generate a list that indicates what sequence of subtasks was performed, the length of time spent on each one, and the frequency with which each of these tasks was performed.
- At the end of the task analysis, the individual responsible for producing a furniture specification will have an accurate list of requirements for the individual and their work.
- Suppliers of workstation furniture should provide a list of suggested workstation furniture that matches the needs of the individual and the task demands.
- Once the shortlist is drawn up, the trial of the workstation furniture can be arranged.
- A trial allows the testing of workstation equipment under controlled conditions.
- The conditions under which a trial is run should be as realistic as possible.
- Individuals who participate in a trial should be representative of those who are likely to use the equipment in the future. The more people who participate in the trial, the better.
- During the trial, participants should perform the full range of tasks they are likely to perform at this workstation furniture should it be introduced in the future.
- If the user is trying out several chair and desk combinations, it is helpful to schedule the changeover between combinations to fit in with natural breaks in the work. This will minimise disruption.
- The trial manager is required to ensure that the trial runs smoothly. He needs to be a good communicator and be able to build a good rapport with the users quickly.

- Feedback needs to be collected from the users during the trials. This can be done using a range of questionnaires, rating scales, or interviews.
- If interviews are used as a means of collecting feedback, trial managers should be capable of subtly steering trial participants so that all of the relevant information is collected.
- Questionnaires or feedback forms can include open-ended or closed questions. Open-ended questions do not limit the answers that can be given. These are more time-consuming to analyse following the trial.
- Rating scales and ranking scales can also be used effectively.
- The environment in which a trial is carried out should be similar to the environment in which the equipment will be used: features such as lighting, temperature, humidity, and noise level should be similar.
- Prior to the roll-out of the equipment selected on the basis of the trial, the organisation should prepare the workforce for the change. Change causes individuals to experience stress because they feel that the situation is out of their control, and this creates feelings of uncertainty.
- Following any changes to workstation equipment, the workforce should be allowed an acclimatisation period in order to get used to the changes before they are expected to work at their normal pace.
- The greater the involvement of the workforce in the trial and implementation phase, the more likely the project will be successful.
- Communication is the most effective means of ensuring that the workforce understands what is intended through the trial process and during the roll-out of new workstation equipment. This approach is likely to overcome resistance.

4

Computer Use

4.1 Introduction

The world of computers has changed radically over the last decade. Computers are no longer seen as pieces of equipment that belong only in an office setting to be used for work purposes. Computers are used in every environment we might encounter during the course of a day and are used in so many diverse ways. For instance, clothing retailers have equipped shop floor staff with tablets so they can carry out on-the-spot checks to see if a particular item is in stock in their store or, if not, whether it can be ordered online. Tablets are used on the streets by people canvassing opinions from passers-by. In warehouses, picking operators work with ruggedized mobile devices that are either handheld or strapped to their arms (with a scanner attached to their finger) so that they can collect a product from the correct location within the racking system and record that the item has been picked and dispatched.

Mobile computer technology has now penetrated most professional and private arenas, and it has pitched itself so that in addition to being a prominent feature of most people's everyday lives, it is increasingly being made available to, and used by, older and technically inexperienced people. Although there may be a stereotype that suggests older people are unable, or unwilling, to learn new technologies, it has been shown (Arning and Ziefle 2006, 2008) that older users want to become acquainted with modern technologies. This brings with it some additional considerations in terms of accommodating user abilities and expectations. Older users place greater demands on designs, and display characteristics of new applications are typically created without considering the needs of this particular population. Due to profound age-related changes in sensory, physical, psychomotor, and cognitive functioning, electronic displaying of information is a challenging issue (Armbrüster et al. 2007; Oetjen and Ziefle 2007; Ziefle et al. 2007).

Ten years or more ago, many users hailed the increased availability, affordability, and flexibility of laptop computers. Nowadays, the advantages of having access to tablets and notebooks have overshadowed the popularity of laptops. One of the main reasons for this increase in popularity is that a tablet is particularly mobile. However, this specific feature, along with its integrated touchscreen, enables users to adopt postures that are very different to the postures adopted when using typical desktop computers. For instance, it seems that tablet users incline their head and neck to a greater extent than would have been seen previously with desktop computers. In certain situations, tablets can be used in a way that promotes greater neck discomfort than would have been the case even when compared with using a laptop.

Although tablets give a user greater flexibility over how and where they use these handheld devices, the device itself offers little or no adjustment as the keyboard and screen are presented in an integrated unit. As a consequence, there appears to have been a trade-off between portability and usability, and this has significance for the user. It also has significance for those who have responsibility for providing computing devices to their workforce. They should keep in mind that any decreases in computer size will have been achieved by changes in hardware design features, such as smaller key sizes, the provision of virtual keyboards, smaller screens, and/or different input methods. These changes could, potentially, impose different demands on the user that, ultimately, could result in feelings of ill health.

Nowadays, businesses commonly assume their workforce is safe and happy using phablets, which straddle the size format of smartphones and tablets. Because they have big displays, they lend themselves to screen-based activity, such as web browsing and multimedia interaction. They may also include software to enable notetaking. However, questions need to be asked about the circumstances in which these devices, and others, are used. Chapter 13 contains a specific checklist relating to mobile devices to assist in gathering this information.

4.2 What's on Offer

There is a dizzying array of choices when it comes to selecting equipment to enable workers to carry out their jobs. If there is any possibility that an individual might work outside of the main office, they are often provided with a mobile PC, and this could take the form of a notebook or tablet. However, that does not provide a true insight into what might be used during the course of a working day and what sort of stresses they might place on the user.

4.2.1 Microcomputers

Microcomputers, which are smaller and lighter versions of laptops, and tablets, such as iPads, may impose their own very particular demands on the user. This will be due to the fact that having been scaled down, the user is presented with smaller keys and monitor, or an integrated monitor/keyboard (which is different from a connected keyboard and screen). This is also likely to be accompanied by changes in input methods. So, instead of hitting keys, the user may have to swipe or gesture or use a virtual keyboard.

4.2.2 Netbooks

Netbooks are very small, thin laptop computers that tend to be lighter than a laptop and less expensive. In contrast to laptops, netbooks are heavily reliant on software that is accessed on the internet rather than on the device itself. The intention behind this feature is to simultaneously reduce the hardware specification of the equipment and their associated costs without undermining the performance of the device. Clearly, netbooks are not designed to be used for tasks that require significant computation.

4.2.3 Ultraportable Laptops

Ultraportable laptops are very lightweight and tend to be under 2kg (just over 4lbs). From a manual handling perspective, this is an important consideration when equipment is being selected, particularly for individuals who are known to travel frequently and who will be carrying their equipment regularly. The ultra-portable is quite compact, offering a screen width of less than 14″ when measured diagonally. Clearly, a consequence of them being scaled down is that they present smaller keyboards and smaller screens when compared with a standard laptop. As these devices are well-suited to word processing, emailing, and surfing, they are a useful work tool. One major plus-point of an ultraportable is that it has an extended battery life, which usually exceeds six hours. This means that it will not always be necessary to take a charging unit along with the device, which will reduce the overall weight being handled.

4.2.4 Laptops

A midsized laptop will have a screen which is 14″–16″ wide. They vary in size depending on the model, and so their weight can range from around 2 to 3.6 kg (4.5 and 8 lbs). These types of laptops are generally used in either the home or office. They offer a suitable range of features and can support most applications that would be required for work purposes, including email, surfing/researching on the internet, photo editing, and multimedia production.

4.2.5 Tablets

The main distinguishing features of tablets are their mobility and touchscreens. Tablets can be selected in two different varieties: slates and convertibles. Convertibles provide built-in keyboards or mice and a pivoting touchscreen. Slates are favoured in situations where the use of keyboards and mice are difficult, such as in hospitals, shops, and restaurants.

4.2.6 Towers

Tower computers are normally associated with work that is carried out in standard offices. They vary in size from cabin baggage size (i.e. a full-size tower) to shoebox size (i.e. a mini- or micro-tower). Towers take the form suggested by their name and stand vertically, usually, on the floor under the desk. Care needs to be taken when positioning them so that they do not hinder the freedom of movement of the seated user.

4.2.7 Desktops

Desktops are similar in size to tower computers and are located in a position suggested by their title: on the desk top. More often than not, the screen is positioned on top. This subject has been dealt with in Chapters 1 and 2.

4.2.8 SFF Computers

Small form factor (SFF) computers are very small desktop PCs that are usually located on the desk surface. They range in size from a shoebox to a paperback book. Their size limits their flexibility in terms of use.

4.2.9 All-in-One Computers

All-in-one computers are desktop units where everything except the keys and mouse is located within the housing of a flat screen. They are particularly suited to multimedia and graphic arts production or 3D modelling and rendering.

4.3 How Desktop Screens Should Be Used

It is common for people who work in offices to blame their computers for being the source of all of their complaints, such as headaches, eye fatigue, and tiredness. Yet, the truth is that these individuals are often the source of their own problems through poor working practices, such as working for long periods without a rest break or change in activity. Despite the fact that they may be the instigators of their own difficulties, they should not necessarily be criticised for working in an inappropriate way. Poor working practices often develop as a result of a lack of training (or a lack of effective training) or, possibly, from a management style that encourages or even coerces people to remain at their desks working without interruption for longer periods than would be recommended. In the main, if some basic principles are applied to the use of a computer, whatever form it takes, most individuals should be able to work without experiencing any adverse effects.

4.3.1 Screens

Desktop screens have grown in size, reduced in size, then grown again over the years. The more recent change in size to larger screens has been driven by the desire to present and view more and more data and applications simultaneously. Many organisations provide their workforce with more than one screen to use for the same reasons.

As screen size grows, so, too, does the amount of desk surface needed to accommodate it. Some desks are too small to allow flexibility in the positioning of the screen, particularly if there is more than one. The knock-on effect of this is that users adopt postures that are dictated by the layout of the equipment on the desk surface, which itself could be driven by the size of the screen.

In a more 'traditional' office arrangement, where the user is working with a single screen, their comfort levels can be compromised by the height of the screen above the work surface. It is common for the screen to be positioned either on top of the desktop computer or directly on the desk surface. Placing a screen on the desk surface can result in taller people having to look down towards the screen, thereby experiencing neck pain. Smaller people often have to look upward when viewing a large screen located on top of a desktop computer. Chapters 1 and 2 offer an overview of how equipment should be arranged to provide a comfortable working posture.

Having established an appropriate position for the screen, users should be aware of the impact relating to the extent of its use. This aspect will be covered in detail in Chapter 7. In brief, it is generally advised that screen-based work should be interrupted at least every hour for about five to ten minutes by either a rest break or a complete change in activity, for example, by attending a meeting. Companies should be aware that accessing the Internet or the company's intranet during break times should be viewed as continuing the screen-based work and not interrupting it.

4.4 Small(er) Computers

The design, development, and introduction of laptop computers, combined with the accessibility of wireless technology, started a major change in the way people could perform screen-based work. They offered freedom over where and when an individual could work. Initially, laptops were seen as providing an opportunity for people to work when away from the office, such as if they were on a train, in a hotel, during a conference, or at home. However, very quickly, laptops were brought into the office more and more, and people started to work with them all day in preference to using their standard desktop equipment. Given the disparity between how the laptop has been designed and how they are actually used, users rapidly started to complain of discomfort associated with their frequent or extended use.

Laptops are an extremely useful device. However, the difficulty associated with them lies in the fact that users are not particularly aware of the features of the laptop that have the potential to create problems for them. They also have little understanding of how such equipment should be used if they want to avoid adverse effects. Equally, some of the environments in which they are used are not conducive to working in a comfortable position. For instance, some users sit on a bed in a hotel room when using their laptops because the room may not provide a suitable alternative. Working in such 'atypical' environments will be discussed in Chapter 5.

If laptops are positioned where the name would suggest—on the lap—head, neck, and wrist postures are more likely to be non-neutral, when compared with positioning it on a desk. This would have the potential to result in increased risk of injury to these areas (Asundi et al. 2010). Despite the potential impact on comfort levels, it has been shown (Moffet et al. 2002) that actual performance was not affected by locating a laptop on the lap, when compared with the desktop. Mobile devices with larger touchscreen sizes, such as tablets, are more likely to be used on the lap due to their increased weight, when compared with smaller devices, such as mobile phones (Kietrys et al. 2015). For that reason, it has been established (Young et al. 2012; Werth and Reeves 2014) that tablets result in greater neck flexion than other portable computing devices. It has also been shown that more awkward wrist postures are likely to be adopted by the hand using the virtual keyboard and by the hand supporting and tilting the tablet when a tablet is placed on the lap (Young et al. 2013).

The forward head posture associated with tablet use can increase cervical extensor strain, leading to fatigue of the extensor muscles and, possibly, the development of neck pain over long-term use (Straker et al. 2009). This is supported by findings that tablet usage in positions requiring greater than neutral head flexion resulted in an increase in the gravitational moment produced by the weight of the head (Straker et al. 2009; Vasavada et al. 2015). The mechanical demand on the neck muscles has been estimated to increase 3–5 times during tablet use, when seated, compared to a neutral seated posture (Vasavada 2015). As gender differences exist in head mass and neck muscle strength, even when users are matched for height, and as there are established gender differences in complaints relating to musculoskeletal disorders involving the neck, this might suggest that gravitational demand may be different for male and female subjects.

Although some users might attempt to remedy the situation of tilting their head forward to look at the tablet by adopting a semi-inclined position, which has the potential to reduce the gravitational effect produced by the head being suspended in front of the shoulders and the subsequent muscle activity of the cervical extensors, there is still a potential for this posture to create increased strain on the passive tissues of the cervical spine (Douglas and Gallagher 2017). As a result, people should adopt the semi-reclined posture with caution, and find something in between the extremes of a very reclined position and sitting upright with the tablet on their lap.

If employees have to work with tablets, such as in meetings, they should place their tablet in a case that allows it to stand up so that they are not trying to look at it when it is flat on a table. Ideally, any inputting should be done with a separate, Bluetooth keyboard. In the alternative, using a stylus on the propped up tablet will allow for a change of hand and wrist posture. Of course, users could always resort to the old fashioned method of pen and paper, and type up their notes back at their desk when they are able to use a standard desktop, which will allow them to adopt a less stressful posture.

In terms of performance, participants in the study by Werth and Reeves (2014) were able to type almost four times as much on the laptop or netbook than on the tablet. As a consequence of both of these findings, the authors concluded that although tablet/slate computers are suitable for gaming or other entertainment uses, they are not suitable for long-term usage as a device for traditional computer work activities. This has implications for business.

Typical laptops are designed with the screen hinged to a keyboard that is non-detachable, and the result has been awkward or constrained body postures and movements (Berkhouta et al. 2004). In particular, studies have shown an increase in neck flexion, torque, and physical discomfort in the neck area. It would appear that the neck muscle load (EMG value) when using a laptop is significantly greater than when using a standard desktop PC. In addition, there is greater forward head inclination, less head movement overall, as well as a reduced viewing distance. As a consequence of the latter, it has been suggested that prolonged laptop use is likely to lead to visual disorders (Jonai et al. 2002). It has been suggested that working with the head bent forward due to the low-level screen is compounded by the fact that laptop screens typically present a quality image to the user only within a narrow range of viewing angles. Therefore, the user is discouraged from changing head position.

The dimensions of the laptop keyboard are usually smaller than those of a standard keyboard. A standard PC keyboard has key spaces of about 19.05 mm; yet, it would not be unusual, by comparison, for a laptop keyboard to have key spaces of 15 mm. Smaller keyboards, such as those on notebooks, are also more likely to result in constrained arm postures and the need for more precise keying (Pheasant and Haslegrave 2006). A study by Lai and Wu (2014) found that the closer the touchpad/touchscreen/keyboard sizes were to a regular sized keyboard, the shorter the operation times. This is supported by the findings of Moffet et al. (2002), who established that compact or virtual keyboards may have significant performance decrements. These findings might suggest that smaller devices could result in slower processing time when working, and this would be likely to have an impact on business efficiency.

As screen height and keyboard dimensions have been identified as being problematic, these aspects have to be tackled if extended use is envisaged. The aim should be to enable the user to adopt an appropriate posture, such as that outlined in Chapter 1. There are two easy ways of tackling this: either by providing a docking station or by providing an external keyboard and mouse. If using a docking station, the user places the laptop into a housing that connects the laptop directly to an external screen, keyboard, and mouse. Having done so, the user can adjust their sitting position relative to the screen and keyboard using an adjustable chair. The alternative is simply to attach an external mouse and keyboard to the laptop, and then raise the laptop up on a platform so that the screen is presented at a suitable height. Again, users should adjust their posture so that they are sitting comfortably. It would be expected that working in this manner will lead to a reduction in mechanical load on the neck in conjunction with an improvement in productivity and reported comfort levels.

In relation to the issue of providing a separate keyboard, it has been reported (Kim et al. 2014) that lower muscle activity is required in finger muscles when typing on a virtual touchscreen, when compared to a conventional keyboard. Werth and Babski-Reeves (2014) also reported lower muscle forces when typing on a virtual (touchscreen) keyboard.

Separate mice are recommended for use with laptops given the poor design of a number of laptop cursor controls. Some have a pointing stick and others have touchpads, which tend to be operated by one fingertip in a bid to avoid sending out 'confusing' commands. In contrast, a standard mouse shares the workload out over a larger portion of the hand, which will be less fatiguing for the limb concerned.

Some office furniture manufacturers provide stands on which the whole laptop can be located so that the screen is raised. However, raising the screen when it is still attached to the keyboard simply raises the keyboard, causing some naive users to work with their arms raised as they depress the keys, unless they have been specifically advised to use a peripheral keyboard. Jacobs et al. (2009) have found that external accessories (such as a laptop/notebook riser, external keyboard, and mouse) and ergonomics training appear to contribute to a trend of decreased self-reported musculoskeletal discomfort, which would suggest they are an option well worth considering. The recommendations outlined above are particularly useful at a semi-permanent workstation such as in an office base or home office. However, using a number of peripherals has limited portability (Asundi et al. 2012), and that may limit the extent of their use in other, more transient, working environments.

The techniques that should be employed when using a standard desktop should also be employed when using a laptop. Once the user has established an appropriate working position, the laptop should then be used in the same manner that a desktop PC would be used in terms of rest breaks and changes in activity. In addition, it should not be assumed that users will see the benefits to be gained from using a docking station or from using an external keyboard and mouse. Some might complain that this causes a degree of inconvenience. Appropriate training and advice should be offered to users so that they appreciate the need to use laptops in conjunction with other equipment.

Following the difficulties related to the height of the laptop screen, the distance between the screen and the seated user can be a source of problems. This results from the need to locate the keyboard at a suitable distance for comfortable use, which, ultimately, brings the conjoined screen closer than an individual might ordinarily prefer. Alternatively, the user pushes the keyboard away to position the screen at a more acceptable distance, which causes them to reach and lean forward when using the keys.

Given the fact that laptops are designed to be 'portable' equipment, they will be carried from one place to another by the users. That being the case, the manual handling element of transporting the laptops becomes important. Anyone who routinely carries a laptop from one work area to another, whether they leave the employer's site or not, should be provided with manual handling instruction that focusses specifically on the safe handling of this type of equipment. Manual handling will be addressed in Chapter 10. Those responsible for selecting the portables that will be used by other employees should be alert to the fact that some of them are heavy and should foresee the potential impact on the individual from carrying such heavy weights. Obviously, purchasing lighter portables should be a primary goal.

Aiming to reduce the weight of the laptop should coincide with a desire to purchase the lightest case in which to carry it. In addition, having the longest battery life possible will reduce the need to carry the transformer every time the laptop is taken somewhere. Alternatively, if users transport the laptop between the office and home, they could be provided with two transformers, one for use at each base, eliminating the need to carry this device between both sites repeatedly. An effort should be made to minimise the number of additional items that might be carried with the equipment, such as documentation.

4.4.1 Mobile Phones/Smartphones

Of course, an assumption can never be made that users will only interact with desktops, laptops, or tablets during the course of their work. Many users, particularly those on the move, will rely on handheld devices, such as their mobile phones, to send and receive emails, search for information on the internet, and so on. It would seem that people who own both smartphones and tablets do not spend less time using their tablets, indicating that smartphones complement tablets and increase total time spent on mobile devices and does not take away from it.

There is a danger, particularly in terms of smartphones, of users allowing a 24/7 'work-aware' approach to creep into their life as a result of linking work email accounts to their personal phones. In effect, they never get to switch off from work mode. This could result in users feeling overloaded and lacking control over work-life boundaries. This is something that businesses need to address by setting out what their expectations are so that the user can justifiably feel comfortable about ignoring an out-of-hours email.

There is significant information available now to suggest that there are potential risks to musculoskeletal health as a result of using mobile devices (Jonsson et al. 2007; Sengupta et al. 2007; Gold et al. 2012). For example, texting or typing on mobile phones requires muscle activity to both hold the device and type. The forces generated through key activation (usually by the thumb) must be counterbalanced and stabilised through activation of the finger flexors and wrist extensors (Kietrys et al. 2015), which increases the workload for the musculature. Encouraging users to employ two-handed grips when reading or browsing through hardware or software design may increase performance and reduce musculoskeletal strain during mobile device use, when compared with a one-handed grip (Trudeau et al. 2016).

Excessive texting on a mobile phone has been associated (Williams and Kennedy 2011) with musculoskeletal disorders in the forearm and thumb, such as tendonitis, tenosynovitis, and first carpometacarpal (CMC) arthritis. It would appear that if mobile device users complain about discomfort, they are most likely to complain about neck discomfort (Xie et al. 2017). It has also been established that head flexion angle was larger when text messaging compared with web browsing and video watching (Lee et al. 2015). It is considered that sustained neck flexion may be a risk factor for developing pain in the neck, shoulder, and upper extremities, and those users who complain of musculoskeletal symptoms are more likely to sit with their head bent forward while texting (Gustafsson et al. 2011).

Most handheld devices require the user to look sharply downwards or to hold their arms out in front of them as they read the screen, which could lead to fatigue and pain in the neck and shoulders (Berolo et al. 2011). It would appear (Gold et al. 2009) that there is a correlation between shoulder and neck discomfort and the number of text messages sent daily. Employees should be advised to raise the phone upwards to keep the head and neck in a more neutral posture. If they are sitting at a desk, they should support their arms on their elbows. However, this should not be a high-frequency practice, as the elbows may become uncomfortable if resting on a hard surface for long periods of time.

In terms of performance, Cannon et al. (2015) established that when inputting text and data using physical keyboards and touchscreen devices, the touchscreen proved to be the optimal handheld device to minimise user error. However, in terms of employee well-being, there would appear to be a link between high keystroke rates and hand disorders, specifically, De Quervain's tenosynovitis and osteoarthritis of the joint at the base of the thumb (Ming et al. 2006; Storr et al. 2007). It would appear that as a result of the small spacing on the mini-keyboard, a greater strain may be placed on the hand and arm muscles when compared with desktop or laptop use (Sengupta et al. 2007). Studies have demonstrated that if the thumb is used for a task such as text messaging, it is located towards the end of its range of movement (Jonsson et al. 2007), and this is likely to increase the loading placed on the extrinsic and intrinsic musculature of the thumb. It is recommended that employees are advised against using their thumb to input text, unless they use both thumbs. It would be less stressful for them to hold the phone in one hand and input text using the forefinger of the other hand. They should also be aware of the advantages of not supporting the bottom of the phone with their finger(s) when using one-handed gripping, as this significantly alters the gripping posture of the hand, making it a more stressful one.

Given the trend towards increasing the size of mobile phones, consideration has to be given to the dimensions of the phone and its relation to the likely hand size of the intended user population. Reference to anthropometric data i.e. body size data, will inform an organisation whether the phone being selected can be held and operated equally well by both males and females given that females would be expected to have smaller hands/fingers.

If a business is aware that employees are frequently sending lengthy texts using a smartphone, it might be beneficial to suggest to them that they try recording and sending voice clips instead.

4.5 Display Characteristics

Because graphical user interfaces (GUIs) are indispensable in various applications and operating systems, the usability of GUIs is increasingly important (Lee et al. 2012). Therefore, there are a number of basic requirements if designers want to ensure that they present information in a meaningful and usable way. If they fail to provide a design and layout that meets human needs, they are likely to have an adverse impact on the ease, accuracy, and efficiency with which users can perform their work.

Note that the following discussion on display characteristics is intended for use with screen displays on desktops, laptops, and other mobile devices. It is not intended for direct application to situations were heads up displays (HUD) and head mounted displays (HMD) are being used.

A basic starting point is that the text should be legible for all age groups. One of the biggest challenges is finding the optimal way of presenting information on a device with a small display. Ziefle (2010) has considered, in particular, how the issue of an increasing number of older users should be assisted when using handheld devices such as mobile phones. Clearly, the displayed information should be easily readable, which means a sufficiently large font size. It was established that navigation performance was optimal when font size and the size of the preview (i.e. the amount of information presented on the screen simultaneously) were large, as the latter enabled more choices to be considered, thereby avoiding unsuccessful choices when navigating.

Generally, users should be able to read the text displayed on any screen and understand it easily, so the text should be in simple language, and they should be able to do this without experiencing any additional visual interference from surrounding material on the screen. From an efficiency perspective, the clearer the image, the more quickly it will be recognised. Legibility is affected by the size of the characters used, the height-to-width ratio, character contrast, clarity, and character format. If the characters on the screen are sharply defined, they will be much easier to read and, as a consequence, will promote a more successful reading performance and reduce visual discomfort. Users will also read the display on a screen more easily and effectively if the characters are familiar to them.

4.5.1 Fonts

A font is a typeface of a specific size and style. Different fonts are used to draw attention to particular areas of a display and differentiate them from others. Although there is a large choice when selecting typeface and character sizes, the aim should be to limit the variation in both to avoid screen clutter, which makes the information content more difficult to decipher correctly. As a general rule, the simpler the font type, the easier the text will be to read. For instance, Latha is easier to read than French Script.

If several styles are combined, the selection should be based on a number of important characteristics. It is easier to read a combination of 'families' if they have similarities, such as the same line thickness, identical heights of the letters, and so on. The term 'family' relates to the variety of forms in which the specific typeface can be displayed. For instance, a particular typeface can be displayed in roman, italics, bold, outline, or shadowed. Having more than two families in a display reduces readability.

Serifs are the cross strokes used on some letters such as Times New Roman. In poor viewing conditions, they can become less distinct, thereby reducing readability. It is usually recommended that sans serif typefaces be used if the type size is 10 point or smaller, if the environment in which the screen is being used is not ideal, or if the screen is of a low resolution. The between character spacing for sans serifs should be a minimum of one stroke width, or one pixel, to optimise readability. If a serif typeface is used, the character spacing should be a minimum of one pixel between the serifs of adjoining characters to optimise readability.

Sans serifs are grouped together to form what is referred to as a 'race'. Roman, including the Times typeface, is another race. The most effective designs include typefaces from one race only. Some designers do mix races, but manage to create an effective display by ensuring that one race is dominant and by assigning each race to a particular purpose on the screen—for example, one style for headings and another for text.

When referring to point size, this offers a description of the distance between the top of the character's ascender and the bottom of its descender. Ascenders are parts of the letter which extend above the main body of a word, and descenders are the parts of the letter which extend below the main body of the word. For instance, in the word 'display' the upper part of the 'd' and 'l' are the ascenders, and the lower part of the 'p' and 'y' are the descenders. Dropping below 8 point is considered to significantly reduce the readability of the screen. To increase readability, the variety of font sizes used should be limited. This is important if families, or races, are being mixed, because one type style within a race may appear to be different in size from another type style within the same race, even if both are the same point size; for example, Latha 12 point appears larger than Times Roman 12 point.

Character heights, i.e. the distance between the top and bottom of a capital letter (without an accent), subtending 20′ to 22′ of arc are considered suitable for most tasks. This describes the angle formed at the eye as a consequence of viewing distance and character height. In some cases, there is no need to read some of the incidental text on the screen. Examples include footnotes and subscripts. In these cases, smaller character heights, such as 16′ of arc, are acceptable. Stroke width, i.e. the distance between the two outer edges of a character stroke, such as across the horizontal line through the letter 't', should be between 1/6 and 1/12 of the character height. For optimum legibility and readability, a character width-to-height ratio between 0.7:1 and 0.9:1 is recommended.

Fonts can also be differentiated by style, such as italics, outlines, and shadows. Italic type is more readable if presented in a serif font and is most effective if used only for small areas of text rather than large sections. Outlined and shadowed effects reduce the ease with which text can be read, so their use should be limited to special graphics. Boldface is commonly used to highlight something important. It is most effective when presented in the sans serif format and used sparingly. Underlining was formerly used to draw attention to important text, but underlining is now associated with the designation of a string of words as a navigation link.

Display designers must choose whether to use upper, lower, or mixed case letters. 'Mixed case' usually means that the designer has followed the format that would be adopted in a handwritten letter: a capital letter at the start of a sentence and lower case for the remaining letters in the sentence, except in acronyms or proper nouns. Using mixed case increases the speed with which an individual can read longer text displayed on the screen. One reason for this is that it gives each word a more distinctive shape, which aids our identification of that particular word. Successful reading depends not only on being able to read each letter in a word, but on recognising the shape of the whole word. The shape gives a degree of 'uniqueness' formed from the different sizes of the letters which makes the word distinct from others and more immediately recognisable. If all uppercase is used, the word loses its unique shape because all of the letters become equal in size, making it less distinctive.

In the case of headings, such as menu choice descriptions, where only single words or short phrases are employed, all uppercase appears to work successfully, particularly as it makes these aspects of the display more prominent. This has significance for displays where warning messages are being displayed. Uppercase can also be used successfully in place of mixed case when the characters would otherwise be unreadable, owing to reduced size.

Individual words are more distinctive if they have a minimum of one character width of space between them. Serifs should not be viewed as contributing to the required degree of spacing between words. Individual lines of text are easier to read if there is a minimum of one pixel of space between each line.

Whatever typeface, font size, family, race, or case is used, the designer should apply it consistently. This will ensure that users can work effectively throughout the whole of the system in use.

4.5.2 Using Colour

When describing a colour, we are actually describing three things: (1) the hue, which is the global meaning of the colour, such as green or red; (2) the saturation, which refers to the purity of the colour when measured along the length of a scale on which one end is a grey version of the colour and the other end is the most vivid version; and (3) the intensity, which refers to the lightness or darkness of the colour and is sometimes referred to as the value. Both psychological and physical factors have a significant impact on the readability of a display. Although the number of measurable colours is considered to be about 7.5 million, it is believed that humans can effectively distinguish only a small portion of these. Therefore, going overboard when using colour might, ultimately, be a waste of time and may interfere with the ease with which displays can be read.

Colour adds interest to the display and makes it more interesting to work with. It can also be used to differentiate between separate pieces of information, particularly when spacing cannot be used instead, and it can be used to identify relationships between items. Colours that are identified as being close together on the spectrum, such as blue and green, are more likely to show a relationship between items displayed on the screen than colours that are widely separated in the spectrum, such as red and blue. The latter are more likely to draw distinctions between displayed items. Colour can also draw attention to important messages rather than using highlighting. It can be used to identify a message's status so that the reader recognises its importance or category without actually reading the content. These two latter points illustrate the use of colour as a formatting tool, or as a means of coding.

The choices of colours are many but only a fraction of what is possible is actually considered effective and usable. If not used advisedly, colour can be distracting and interfere with the usability of the system. Great thought has to be given to which colours should be employed in a display design and what combinations should be used. If the wrong colour choices are employed, there can be a deterioration in performance as a direct result of unintended visual effects. Colours can be considered to have been used successfully if users are able to detect the colour against a visually noisy background, if they can identify it correctly, if they can discriminate between colours, and if they associate the correct meaning with the colour. It is generally believed that no more than 11 colours should be used simultaneously in one display, if accurate identification is required. The general rule is to minimise the number of colours used where possible. The more colours used, the longer the response time and the more likely it is that the user will make mistakes owing to confusion. The way to control the number of colours used is to avoid using colour when other methods of identifying information are available, such as positioning on the screen or highlighting. If it is intended that the user will perform a rapid visual search of the display, no more than six colours should be employed. Any more than this will result in a reduction in comparative discrimination—the ability to identify a colour correctly while other colours are displayed alongside it on a screen. More than six will also have an adverse effect on recalling what the colour means. This suggests that extra thought needs to be given to situations where users can customise their own display colours. A limit may need to be placed on the number of colours they can choose and the combinations they use.

Colours widely spaced on the colour spectrum promote good comparative discrimination. It should be kept in mind that if blue is used, numbers or small icons may not be legible. Blue is, however, considered a good background colour.

Using more than four or five colours will result in a deterioration of absolute discrimination, in which a colour is correctly identified when no other colours are displayed alongside it on the screen. In this instance, the colours should be widely spaced on the spectrum (e.g. red, yellow, green, blue, brown). Using a key to illustrate the colours and to describe their associated meanings could be useful with this type of discrimination. Brighter colours are likely to be more effective for extended viewing, for emphasising information, and for older viewers. The following colours are listed in descending order of perceived brightness: white, yellow, green, blue, red. If there is a need to emphasise a distinction between types of information, then contrasting colours should be used. If the intention is to emphasise the relatedness of information, then similar colours should be used, for example, blue and green. It should be kept in mind that high levels of lighting in the viewing environment will have an impact on the perceived brightness of colours. Colours may appear to be washed out in bright light, and this may result in more errors in identifying the colours. Therefore, the colour chosen for the display may need to be determined to an extent by the environment in which it is to be used. On that basis, if trials are being arranged to test the likely success of a system, they should be run in an environment where the lighting level is the same as that where it will ultimately be used.

The perceived appearance of colours is influenced by other factors, such as the total size of the area covered by the colour in question, the other colours that fall within the viewing area, the task being performed, and any colour-viewing deficiencies in the user group. Colour-viewing deficiencies are referred to more commonly as 'colour blindness'. This is not a particularly accurate term, because the individuals in question are not blind to a particular colour, they simply have difficulty in discriminating between specific colours. Protanopia is a deficiency in viewing red; deuteranopia is a deficiency in viewing green; and tritanopia is a deficiency in viewing blue. Individuals with these types of deficiencies may not be able to perceive a discernible difference between hues, but they are likely to perceive a difference in lightness or intensity. Even with these deficiencies, thoughtful display design will ensure that these individuals can still use a system effectively.

A user's understanding of what a colour symbolises may have an impact on how successfully it can be used. Colours have different meanings in different contexts, and if they are used out of context, they could cause confusion or interference. For instance, applying green to a display command that is intended to stop the user may not achieve the desired result given its association for many with 'go'. It is also considered to be the case that the degree of importance assigned to a word can be influenced by the colour in which it is presented. This has implications in terms of warning compliance. It is generally accepted that red, orange, yellow, and white are interpreted by readers to represent decreasing levels of hazard.

Cultural stereotypes also influence the success of a user's interaction when viewing colours on a screen display. For instance, the colour red is often associated with danger in the UK and United States. However, in Japan, it is associated with happiness. Inappropriate use of a colour in a particular culture could create problems in terms of performance. This underlines the importance of generating a full understanding of a potential user's typical experiences and expectations, both of which will affect the success of the display design.

When actually choosing what colours will be used, the designer has to understand how the information being viewed by the individual will be used. This is essential if the chosen colour is intended to be used to categorise information into organised blocks on the screen. All of the relevant information has to be presented in the appropriate colour for its category.

Colour can be used to categorise information on the screen so that the user can attend to each section successively or to follow a particular sequence; it can be used to categorise information with the same degree of importance; it can be used to categorise information of a particular type, or it can be used to categorise large quantities of information so that they are presented in groups (chunks) making it easier to comprehend.

The positioning of colours on the screen is not a haphazard affair. To work effectively, the use of colours and their location should be influenced by the way in which the eye works. The eye is most likely to register green and red if they are employed in the middle of the visual field. Because the retina is most likely to register blue, white, black, and yellow at its periphery, it is more effective to locate these colours on the outer edges of the visual field. If decisions have to be made about how colours will be arranged when several are being used together, it should be kept in mind that people 'expect' a natural order to colour, and that the most acceptable order follows the colours of the rainbow: red, orange, yellow, green, blue, indigo, and violet. Simultaneous use of highly saturated blue and red will cause repeated refocussing, because the eye can only focus on one of the two extreme wavelengths at a time, and a user is likely to experience fatigue as a result.

Consideration has to be given to the issue of how colours are going to be layered—in other words, what will form the background and what will form the foreground. If the foreground is very different from the background, this will increase legibility.

For highlighting the foreground, a lighter version of the foreground colour should be used. Some designers have chosen to use a darker version of the foreground colour to 'highlight' information, thinking that this will be sufficient to draw attention to that information. However, this approach results in 'lowlighting' as opposed to 'highlighting', which will remove the emphasis from the section in question. If lowlighting is intentional, it should not be used in conjunction with highlighting, because this can confuse the viewer.

The background colour should be chosen carefully so that it functions as a passive means to group related sections of information together. It should not be in competition with the foreground for attention. At the same time, poor contrast between brightness and hue of the background and foreground will reduce legibility (for instance, blue text on a black background).

Colours, of course, are not the only option to generate more successful screen displays. The spatial layout can also be employed effectively. Sometimes, the location of a piece of information is sufficient to draw attention to it, and using colour is unnecessary. Whatever colours are selected, they should be used consistently from screen to screen and across systems.

4.5.3 Layout

Designing a display system that flows logically will increase the speed with which information is processed and the accuracy of the user's response. If designers wish to create a layout that can be used successfully, they need to follow a number of general principles. People from Western cultures tend to start scanning information from the top left, whether the information is presented in printed form, such as a book, or on a screen. Thus, the information that will be used most frequently, or that is most important, should be located in the top left-hand corner of the screen. The most effective layout is one that follows a top-to-bottom, left-to-right format.

The information should be ordered in a logical sequence. It should be prioritised so that the most critical information is presented first, subsequently cascading down to information of the least importance. The sequence should minimise not only movement of the input device when making selections, but also should minimise eye movement. Scrolling should be minimised where possible. The same sequence should be employed within the system and across systems. Ideally, information should be aligned. If manuals are to be used in conjunction with the screen, the order in which the material appears on the screen should match the order in which it is presented in the manual.

When information is presented, it should be of sufficient length to convey the message, but not so long as to be overwhelming. The individual pieces of information required to complete one operation should be presented on a single screen, if possible. If more information is presented on a single screen than is actually required to complete the desired task, resulting in overcrowding, the user will take longer to process all of the information and may make more mistakes. Scrolling down lengthy screen pages should be avoided owing to the fatiguing effect of using the mouse over extended periods.

Related information should be grouped together with sufficient space between groups to show a clear demarcation. If there are a large number of groups of information, then the user should be presented with headings and subheadings to avoid overcrowding. Each heading should be a precise indicator of the types of groups that constitute its subheadings. If users have to transfer from one screen to another to complete a task, they should not be required to memorise information from one screen for use on another.

When navigating around the system, the extent to which the mouse and keyboard are used should be kept to a minimum, and the number of times the user has to transfer the hand between these two items should be limited. Building in short cuts for experts might be helpful; however, it should be kept in mind that using commands rather than working through menus places a greater burden on memory. Response time should be minimised and users should be provided with a message indicating a delay if one occurs.

4.5.4 Wording

Carefully chosen words that convey meaning easily are most likely to maximise the efficiency with which people can use the system. The words employed should be simple and clear, and, to that end, the use of abbreviations or jargon should be avoided. Normal punctuation should also be used. Words that would be used in everyday language by the workers are more readily understood, the emphasis being on familiarity rather than word length. Once words have been selected for inclusion in a screen display, they should be used consistently, and the focus should be on using positive words rather than words with negative overtones. For example, using the sentence 'enter data in all boxes before clicking on send' is more positive than 'do not click on send before entering data in all boxes'. Correcting inappropriate use of the system should also be accomplished using positive words rather than negative words that can be perceived as threatening. For example, rather than informing users that they have made an 'error' when entering data, they should be prompted to 're-enter names in alphabetical order'. This prompt draws the user's attention to the important element of the information and how it should be entered. A balance should be struck so that the wording does not slip from being offensive or accusatory to condescending. It is generally believed that attempting to work any type of humour into a display intended for work use should be avoided.

Verbs that are active rather than passive encourage more efficient use of a system. For instance, 'Press F1 to return to main menu' is more immediately understood than 'The main menu can be accessed by pressing F1'.

Underlining wording as a means to draw attention to it should be used only in situations where the user is not under time pressure to complete a task. Underlining results in a reduction in interline spacing and masking of the lower sections of the characters, thereby making the words more difficult to read. Presenting whole words in a capitalised format also hinders performance.

Overall, the screen display should be visually pleasing. To achieve that, the display should be: balanced, symmetrical, regular, predictable, sequential, economical, unified, proportional, simple, and grouped.

4.5.5 Windows

It is common for users to work with more than one window simultaneously. These can be in a tiled format, where the windows are laid out on the screen like floor tiles positioned side by side, but usually in different sizes or overlapping. What needs to be kept in mind when selecting the style of presentation is that in the overlapping styles, some windows may obscure information on other windows. To access the information on other windows, users may have to click the mouse more frequently to switch between windows. Care should be taken that users do not have to significantly increase the extent to which they use the mouse as a result of the style of presentation employed, as this may increase fatigue and discomfort in the upper limbs.

Occasionally, additional windows are employed to display information, such as scanned documents. These documents can be printed forms or simply handwritten letters that have been sent to a company and have to be processed through the computerised system. Users are required to read the contents of the scanned document and input details into other windows displayed on the screen. The quality of scanned documents will have an impact on the ease with which the material can be read, and care should be taken that users, possibly already completing a demanding screen-based task, should not be faced with the added burden of attempting to decipher poor reproductions. The provision of a 'magnification' tool that can be accessed easily and rapidly would be helpful, but again, using such a tool is likely to result in increased usage of the mouse.

4.6 Summary

- Problems encountered by people who use computers are often a result of working for long hours without a rest break or a change in activity.
- Poor working practices often result from a lack of training or lack of effective training. Management style can also influence the development of poor working practices.
- Screens are frequently placed either directly on the desk surface or on top of the hard drive. This does not allow a sufficiently flexible arrangement to accommodate all people comfortably.
- Screen-based work should be interrupted at least hourly for about five to ten minutes by a rest break or by a change in activity.
- The introduction of laptop computers has transformed the way people perform screen-based work.
- Laptop computers offer freedom over where and when people can work. This has been speeded up with the advent of wireless technology.
- Laptops have become so popular that they now are being used in place of standard screens and keyboards in an office, and their extended use is creating problems for the users. The design of the laptop is likely to cause discomfort if used for extended uninterrupted periods.
- The main problems associated with laptops include a low-level screen that requires the user to look down at a low level. Working with a head bent forward is likely to lead to problems in the neck, if done for extended uninterrupted periods.
- If the laptop keyboard is set at a suitable distance from the user, the screen is presented too close for comfortable viewing. This has been related to the development of visual disorders.

- Laptop keyboards are smaller than standard keyboards, which results in constrained arm postures.
- Laptop pointing sticks or touchpads are not suitable for extensive use.
- The problems associated with laptop use can be dealt with by using either a docking station or by using an external keyboard and mouse.
- Laptops should be used in the same manner as a standard PC in terms of rest breaks and changes of activity.
- When selecting laptops, consideration needs to be given to their weight and the manual handling techniques involved in their movement.
- People carrying laptops as part of their working day should be provided with manual handling training.
- Care should be taken that additional items are not carried unnecessarily along with a laptop, which will increase the total weight being handled.
- The world of computers has changed radically over the last decade.
- Computers are used in every environment we might encounter during the course of a day and are used in so many diverse ways.
- Mobile computer technology has now penetrated most professional and private arenas and has pitched itself into being a prominent feature of most people's everyday lives.
- Older users want to become acquainted with modern technologies.
- Older users place greater demands on designs, and display characteristics of new applications are typically created without considering to the needs of this particular population.
- The advantages of having access to tablets and notebooks has overshadowed the popularity of laptops.
- One reason for this increase in popularity is that a tablet is particularly mobile.
- The mobility of a tablet, along with its integrated touchscreen, enables users to adopt postures that are very different to the postures adopted when using typical desktop computers.
- Tablets can be used in a way that promotes greater neck discomfort than a laptop.
- Portable devices present a trade-off between portability and usability, and this has significance for the user.
- Decreases in computer size will have been achieved by changes in hardware design features. These changes could impose different demands on the user that, ultimately, could result in feelings of ill-health.

- Microcomputers are scaled down and present small keys and monitor, or an integrated monitor/keyboard which can be accompanied by changes in input methods.
- Netbooks reduce the hardware specification of the equipment and their associated costs without undermining the performance of the device. They are not designed to be used for tasks that require significant computation.
- Ultraportable laptops are very lightweight and are scaled down. They have an extended battery life.
- Tablets are mobile and have touchscreens. They are useful in situations where the use of keyboards and mice are difficult.
- Care needs to be taken when positioning towers under a desk so that they do not hinder the freedom of movement of the seated user.
- If laptops are positioned on the lap, head, neck, and wrist postures are more likely to be non-neutral when compared with positioning them on a desk.
- Tablet use results in significantly more non-neutral wrist, elbow, and neck postures.
- Tablets result in greater neck flexion than other portable computing devices.
- Tablet use results in significantly more non-neutral wrist, elbow, and neck postures compared to a desktop.
- Research concludes that tablet/slate computers are suitable for gaming, but are not suitable for long-term usage as a device for traditional computer work activities.
- Performance is not affected by locating a laptop on the lap when compared with the desktop.
- Prolonged laptop use is likely to lead to visual disorders.
- If employees have to work with tablets, they should place their tablet in a case that allows it to stand up so that they are not trying to look at it when it is flat on a table.
- Any inputting on a tablet should be done with a separate Bluetooth keyboard.
- Using a stylus on the propped up tablet will allow for a change of hand and wrist posture.
- There is an increase in the visual and musculoskeletal workload when using a laptop, when compared to a desktop.
- The dimensions of the laptop keyboard are usually smaller than those of a standard keyboard.
- Smaller devices can result in slower processing time.

- There is significant information to suggest that there are potential risks to musculoskeletal health as a result of using mobile devices such as mobile phones.
- Texting/typing on mobile phones requires muscle activity to both hold the device and type.
- The forces generated through key activation (usually by the thumb) must be counterbalanced and stabilised through activation of the finger flexors and wrist extensors which increases the workload for the musculature.
- Excessive texting on a mobile phone has been associated with musculoskeletal disorders in the forearm and thumb.
- Head flexion angle was larger when text messaging compared with web browsing and video watching on mobile phones.
- There is a correlation between shoulder and neck discomfort and the number of text messages sent daily.
- Two-handed grips on mobile phones may increase performance and reduce musculoskeletal strain when compared with a one-handed grip.
- If the thumb is used for a task such as text messaging, it is located towards the end of its range of movement, which will increase the loading placed on the extrinsic and intrinsic musculature of the thumb.
- Employees should be advised to raise the phone upwards to keep the head and neck in a more neutral posture.
- If they are sitting at a desk, they should support their arms on their elbows; but this should not be a high frequency practice, as the elbows may become uncomfortable if resting on a hard surface for long periods or frequently.
- Employees should send voice clips rather than long work-related texts using their mobile phone.
- The manner in which information is presented to the user on a screen display, the combination of colours, and the wording used all determine how accurately and quickly the information is processed.
- Developing a suitable screen display requires an accurate understanding of human memory, perception, information processing, reaction time, and visual capability.
- Well-designed screen displays result in increased user satisfaction, greater usage of the system, improved performance, and productivity.
- The design of a display is more likely to be effective and maintain high performance levels if it is meaningful and usable.

- The text displayed on a screen should be legible by all age groups.
- Users should be able to distinguish the text from the background and surrounding materials easily.
- The text should be easily understood.
- The text should be in simple language.
- The clearer the image, the more easily recognisable the words will be.
- Legibility is affected by the size of the characters used, the height-to-width ratio, character contrast, clarity, and character format.
- A font is a typeface of a specific size. Fonts draw attention to particular areas of a display and differentiate them from each other.
- Typefaces and character sizes should be limited to avoid screen clutter.
- Simple font styles should be selected because they are easy to read.
- Family relates to the various forms in which specific typefaces can be displayed. It is easier to read a combination of families if they share similarities.
- Serifs are cross strokes seen on some letters.
- In poor viewing conditions, serifs become less distinctive, making text less readable.
- As an alternative to serifs, sans serifs should be used.
- The term 'point size' refers to the distance between the top of a character's ascender and the bottom of a character's descender.
- Character height and stroke width have an impact on legibility.
- Italics are most effective if kept to small areas of text only.
- Boldface is most effective when used sparingly.
- Using mixed case letters, which is a combination of upper and lower cases, increases the speed with which people can read longer text on a screen. The combination of upper and lower case gives shape to particular words, which makes them more distinct from others and more immediately recognisable.
- Using uppercase for the whole word results in it losing its shape and being less distinctive.
- Whatever typeface, font size, family, race, or case is used, the screen display should apply the choices consistently.
- Use of colour adds interest to a display, and, if used correctly, can assist in differentiating between separate sections of information. Colour can also be used to identify relationships between items.

- Colour can be used to draw attention to important messages and to identify a message's status so the person does not actually have to read the content.
- If colours are not used advisedly, they can be distracting and interfere with the usability of the system.
- If the wrong colour choices are employed, performance can deteriorate.
- Where possible, the number of colours should be minimised. The more colours that are used, the longer an individual will take to read and understand the material, and users will be more likely to make mistakes.
- The environment in which the colour display is read will have an impact on how easy the material is to read.
- A user's understanding on what a colour symbolises may have an impact on how successfully it can be used. Colours have different meanings in different contexts, and, if they are used out of context, they may cause confusion.
- Cultural stereotypes also influence the success of colour choices in displays.
- Careful consideration has to be given to colour combinations. The choice of foreground and background colours can affect legibility.
- The choice of colours should be used consistently from one screen to another and across systems.
- Appropriate layout of a display will increase the speed with which information is processed and the accuracy of the user's response.
- Information presented on the screen should be ordered in a logical sequence and prioritised so that the most critical information is presented first.
- Scrolling through long lists of information should be avoided.
- Overcrowding of the screen should be avoided because it takes users longer to process all of the information, and they may be more susceptible to making mistakes.
- Overcrowding can be avoided by using headings and subheadings, with each heading being a precise indicator of the types of groups that will constitute the subheadings.
- The wording chosen for inclusion in a display should convey meaning easily.
- The words should be simple and clear, and abbreviations and jargon should be avoided.

- Normal punctuation should be used, and the words should be in everyday language that is readily understood.
- The emphasis should be on using positive words rather than words with a negative overtone.
- Active rather than passive verbs result in more efficient use of the system.
- Underlining should be used sparingly, because it masks the lower section of characters, making them difficult to read.
- The use of overlapping windows should be limited to avoid excessive use of the mouse as the user clicks between different screens.

References

Armbrüster, C., Ziefle, M., and Sutter, C. 2007. Notebook input devices put to an age test: The usability of trackpoint and touchpad for users over 40 years of age. *Ergonomics* 50 (3), 426–445.

Arning, K. and Ziefle, M. 2006. What older adults expect from mobile services. In: Pikaar, R. N., Konigsveld, E. A., and Settels, P. J. (Eds.), *Proceedings of the IEA 2006: Meeting Diversity in Ergonomics*. Amsterdam, the Netherlands: Elsevier.

Arning, K. and Ziefle, M. 2008. Assessing computer experience in older adults: Development and validation of a computer expertise questionnaire for older adults. *Behaviour & Information Technology* 27 (1), 89–93.

Asundi, K., Odell, D., Luce, A., and Dennerlein, J. T. 2010. Notebook computer use on a desk, lap and lap support: Effects on posture and comfort. *Ergonomics* 53, 74–82.

Asundi, K., Odell, D., Luce, A., and Dennerlein, J. T. 2012. Changes in posture through the use of simple inclines with notebook computers placed on a standard desk. *Applied Ergonomics* 43, 400–407.

Berkhouta, A. L., Hendriksson-Larsen, K., and Bongers, P. 2004. The effect of using a laptop station compared to using a standard laptop PC on the cervical spine torque, perceived strain and productivity. *Applied Ergonomics* 35 (2004), 147–152.

Berolo, S., Wells, R. P., and Amick, B. C. III. Musculoskeletal symptoms among mobile hand-held device users and their relationship to device use: A preliminary study in a Canadian university population. *Applied Ergonomics* 42 (2011), 371–378.

Cannon, A. B., Strawderman, L., and Burch, R. 2015. Evaluating change in user error when using ruggedized handheld devices. *Applied Ergonomics* 51 (2015), 273–280.

Douglas, E. C. and Gallagher, K. M. 2017. The influence of a semi-reclined seated posture on head and neck kinematics and muscle activity while reading a tablet computer. *Applied Ergonomics* 60 (2017), 342–347.

Gold, J. E., Driban, J. B., Thomas, N., Chakravarty, T., Channell, V., and Komaroff, E. 2012. Postures, typing strategies, and gender differences in mobile device usage: An observational study. *Applied Ergonomics* 43 (2012), 408–412.

Gold, J. E., Kandadai, V., and Hanlon, A. 2009. *Texting and Upper Extremity Symptoms in College Students*. Philadelphia, PA: American Public Health Association Conference.

Gustafsson, E. et al. 2011. Technique, muscle activity and kinematic differences in young adults texting on mobile phones. *Ergonomics* 54 (5), 477–487.

Jacobs, K., Johnson, P., Dennerlein, J. et al. 2009. University students' notebook computer use. *Applied Ergonomics* 40 (2009), 404–409.

Jonai, H., Villanueva, M. B. G., Takata, A., Sotoyama, M., and Saito, S. 2002. Effects of the liquid crystal display tilt angle of a notebook computer on posture, muscle activities and somatic complaints. *International Journal of Industrial Ergonomics* 29, 219–229.

Jonsson, P., Johnson, P. W., and Hagberg, M. 2007. Accuracy and feasibility of using an electrogoniometer for measuring simple thumb movements. *Ergonomics* 50, 647–659.

Kietrys, D. M., Gerg, M. J., Dropkin, J., and Gold, J. E. 2015. Mobile input device type, texting style and screen size influence upper extremity and trapezius muscle activity, and cervical posture while texting. *Applied Ergonomics* 50 (2015), 98–104.

Kim, J. H., Aulck, L., Bartha, M. C., Harper, C. A., and Johnson, P. J. 2014. Differences in typing forces, muscle activity, comfort, and typing performance among virtual, notebook, and desktop keyboards. *Applied Ergonomics* 45, 1406–1413.

Lai, C.-C. and Wu, C.-F. Display and device size effects on the usability of mini-notebooks (netbooks)/ultraportables as small form-factor Mobile PCs. *Applied Ergonomics* 45 (2014), 1106–1115.

Lee, D., Kwon, S., and Chung, M. K. 2012. Effects of user age and target-expansion methods on target-acquisition tasks using a mouse. *Applied Ergonomics* 43 (2012), 166–175.

Lee, S., Kang, H., and Shin, G. 2015. Head flexion angle while using a smartphone. *Ergonomics* 58 (2), 220–226.

Ming, Z., Pietikainen, S., and Hänninen, O. 2006. Excessive texting in pathophysiology of first carpometacarpal joint arthritis. *Pathophysiology* 13, 269–270.

Moffet, H., Hagberg, M., Hansson-Risberg, E., and Karlqvist, L. 2002. In-fluence of laptop computer design and working position on physical exposure variables. *Clinical Biomechanics* 17, 368–375.

Oetjen, S. and Ziefle, M. 2007. The effects of LCD's anisotropy on the visual performance of users of different ages. *Human Factors* 49 (4), 619–627.

Pheasant, S. and Haslegrave, C. 2006. *Bodyspace: Anthropometry, Ergonomics and the Design of Work*. 3rd ed. Boca Raton, FL: CRC Press.

Sengupta, A., Grabiner, S., Kothari, P., and Martinez, G. 2007. Ergonomic aspects of personal digital assistant (PDA) and laptop use. In: *Proceedings of the Sixth International Scientific Conference on Prevention of Work-Related Musculoskeletal Disorders*, Boston, MA, 27–30 August.

Storr, E. F., de Vere Beavis, F. O., and Stringer, M. D. 2007. Texting tenosynovitis. *New Zealand Medical Journal* 120 (1267).

Straker, L., Skoss, R., Burnett, A., and Burgess-Limerick, R. 2009. Effect of visual display height on modelled upper and lower cervical gravitational moment, muscle capacity and relative strain. *Ergonomics* 52 (2), 204–221.

Trudeau, M. B., Asakawa, D. S., Jindrich, D. L., and Dennerlein, J. T. 2016. Two-handed grip on a mobile phone affords greater thumb motor performance, decreased variability, and a more extended thumb posture than a one-handed grip. *Applied Ergonomics* 52 (2016), 24–28.

Vasavada, A. N., Nevins, D. D., Monda, S. M., Hughes, E., and Lin, D. C. 2015. Gravitational demand on the neck musculature during tablet computer use. *Ergonomics* 58 (6), 990–1004.

Werth, A. and Babski-Reeves, K. 2014. Effects of portable computing devices on posture, muscle activation levels and efficiency. *Applied Ergonomics* 45 (2014), 1603–1609.

Williams, I. W. and Kennedy, B. S. 2011. Texting tendinitis in a teenager. *The Journal of Family Practice* 60 (2), 66–68.

Xie, Y., Szeto, G. and Dai, J. 2017. Prevalence and risk factors associated with musculoskeletal complaints among users of mobile handheld devices: A systematic review. *Applied Ergonomics* 59 (2017), 132–142.

Young, J. G., Turdeau, M., Odell, D., Marinelli, K., and Dennerlein, J. T. 2012. Touch-screen Laptop user configurations and case-supported tilt affect head and neck flexion angles. *Work* 41, 81–91.

Young, J. G., Trudeau, M. B., Odell, D., Marinelli, K., and Dennerlein, J. T. 2013. Wrist and shoulder posture and muscle activity during touch-screen tablet use: Effects of usage configuration, tablet type, and interacting hand. *Work* 45, 59–71.

Ziefle, M. 2010. Information presentation in small screen devices: The trade-off between visual density and menu foresight. *Applied Ergonomics* 41 (2010), 719–730.

Ziefle, M., Schroeder, U., Strenk, J., and Michel, T. 2007. How young and older users master the use of hyperlinks in small screen devices. In: *Proceedings of the SIGCHI Conference on Human Factors in Computing Systems 2007*. Association for Computing Machinery, New York, pp. 307–316.

5

'Atypical Offices'

5.1 Introduction

Although the majority of people who interact with computers as a normal part of their working day are located within a central office building, many users are now working from a staggering array of environments outside of their employer's premises. Agile working, remote working, mobile working, working from home–whatever it is that people do, today's work is all about what they do and not where they do it.

There are also a huge number of people working with computers even though they are not being used for typical office-type work for commercial organisations. For instance, in many factories, production lines are monitored and altered with the use of computers positioned alongside the line. Operators working in assembly and manufacturing environments refer to computer screens for instructions on the next phase of their work and will touch the screen to confirm that they have completed a specific operation. These latter environments stretch the ability of the health and safety advisor or human factors consultant to provide a satisfactory working position for the user because the workstation may not consist of a simple desk and chair, but might be constructed of a workbench with jigs, tools, and bulky parts.

The one helpful feature of trying to provide advice in situations that would be considered atypical is that people have basically the same requirements irrespective of the environment in which they are working. Therefore, the human factors specialist is always trying to work towards the same goal, no matter where the users find themselves. The basic, unchanging principles are:

a. The user should be able to remain upright whether seated or standing.
b. The user should be facing forward.
c. The user's head should be relatively upright, but allowed to relax into a slight forward inclination—this will happen naturally due to the weight of the head, unless the screen is too high.

d. The screen should be positioned so that the top of the viewable area does not pass above the user's horizontal line of sight.
e. The area of a touchscreen that needs to be tapped with the finger should not be above shoulder height—this is dictated to an extent by the software design and where the 'touch/next' window is positioned on the display; so, it is worth bearing in mind if software is being commissioned or tailored for the business.
f. If seated, the user should have a comfortable supportive seat or be able to insert an 'add-on' to support their lower back in particular.
g. The user's feet should be firmly on the floor or another support.
h. If a keyboard is being used, the keys should be presented at around elbow height.
i. The keyboard should be within easy reach when there is a 90° angle at the elbow.
j. If a separate mouse is being used, it should be placed on a solid surface at about elbow height.
k. Given the less than ideal circumstances these users are likely to find themselves in, their exposure to this environment should be limited.

Finally, it is worth suggesting that employers should have a realistic expectation of what a user can achieve when working away from the office. Allowances need to be made for the fact that they might be working in far from ideal circumstances, they may need to take more regular breaks, and they may need to limit the amount they try to do.

5.2 Out-of-Office Working

People have been able to work at a distance from their office base for some time. Wireless technology has increased the opportunities for working remotely. Despite the fact that employees might be out of sight, employers still have a responsibility for their well-being. Places where users might typically be found working are at home, in vehicles, in hotel rooms, on trains, and in planes. Heasman et al. (2000) have shown a correlation between working in 'nonstandard' locations, such as at home and on public transport, and discomfort.

5.2.1 Home Office

Many users, when permitted, will work from home occasionally. This may involve merely one hour of work or a whole day. It might be an irregular occurrence or a daily one. Others may spend the whole week, or a major proportion of it, working from home. The more frequently individuals work from home and the longer they spend working at home on each occasion will dictate that an assessment of their environment should be carried out. Many users employ a dining room table or kitchen table with any chair that is available. Some perch on an armchair and place their laptop on a coffee table. Some will be provided with a budget to buy workstation furniture for use in their homes. Unfortunately, many of these individuals will choose workstation furniture that is compatible with the look of the other furniture in their house, particularly if one room has to double-up as a home office during the day.

Working from home has to be carefully controlled, and the same guiding principles that would be applied to a conventional office apply to the home office environment. In other words, people should aim to adopt the appropriate postures while working by arranging their furniture and equipment appropriately. If individuals work from home consistently, say for at least an hour every day, they should be provided with suitable desks and chairs, as well as anything else deemed appropriate, such as a footrest. A home worker should be approached and dealt with in the same manner as an office-based worker if the frequency and duration of the work warrant it.

Even if home workers only perform occasional work at home, and perhaps do not merit the provision of standard office furniture by the company, they should still be offered advice on how to make themselves comfortable whilst working. Whatever surface they are using to support their computer, desktop or portable, users should sit at an appropriate height relative to the keyboard. Sitting on a dining room chair beside a dining room table often results in users working with their arms raised. If a user does not have a height-adjustable chair, they should sit on a cushion to raise themselves. Having done so, if their feet are not touching the floor, they should use some form of foot support. If a footrest is not available, any object that offers a large stable base, such as a ream of copier paper, can be used to support the feet.

Often, dining room and kitchen chairs have solid straight backs, which, obviously, will not offer complete support to a user's contoured back. Sometimes a rolled-up towel placed in the small of the back will make the user more comfortable.

The screen and keyboard of a desktop should be set up properly using the same techniques employed in the main office. If using a laptop, users should be provided with a separate conventional keyboard and mouse. They should place the laptop on a stable raised surface, such as a ream of copier paper or box file, if a proper height-adjustable surface is not available, so that the screen can be viewed at an appropriate height once they are sitting down to work. If users have been provided with conventional office furniture, they must be given a thorough briefing on how the furniture should be used.

For the reasons outlined in Chapter 4, users should not use tablets or mobile phones for lengthy periods of time, as the reduced size of the device, the lack of a separate keyboard, and the predictable location of the device are likely to compromise the comfort of the user.

Home workers require the same detailed information about work patterns and break taking as would be offered to an office-based worker. This will be discussed in Chapter 7. There are some additional issues that need to be kept in mind which are unique to home workers. For instance, although they may work a typical eight-hour day, they may fit this in around domestic arrangements, such as driving children to school. Sometimes parents may then work without a break once they get home, even working through their lunch break to make up the time.

Overall, it is not recommended that users be left to their own devices once they work from home. They need guidance on the most appropriate ways to work, and an assessment needs to be carried out of their workstations.

5.2.2 Transient Workplaces

It is very common for office workers who use laptops and tablets to find themselves in transient environments such as hotels, conferences, and conference rooms. In fact, many of them will work under the same constraints if they go to meeting rooms or break-out areas in their own offices. One of the major advantages of tablets is their mobility, and this has resulted in them being used in preference to laptops when workers telecommute. It is believed that the potential for injury or illness might be elevated as a result of working on smaller, portable computers in non-traditional work settings (Werth and Kari Babski-Reeves 2014).

Users have to exert some control over the environment in which they find themselves when working away from the office and away from home. For instance, it has been found (Asundi et al. 2012) that users of tablets, in particular, will be able to achieve improved head and neck postures by simply increasing the angle of the tablet on a support surface with an improvised method, such as a small ring binder or even the transformer of the power supply tucked under the device. This is assuming that they have not taken a purpose built, lightweight tablet stand with them. However, if the user is actually interacting with the keyboard on the tablet, as opposed to simply reading the display, there is a trade-off in terms of greater wrist extension, i.e. bending the hands up at the wrist. Clearly, using a separate keyboard and mouse will improve the situation, and, given the small, lightweight keyboards currently available, this should not be excluded as a possibility on the basis of having another item of equipment to carry.

Users who find themselves balancing their laptops on their laps habitually as they attend conferences or meetings, or while they wait at the gate in an airport, should look for a means to raise the laptop up so that they do not have to lean forward and downward as they look at the screen and use the keys. The other issue is that for many people their thighs will slope downhill as the seat they are using may not match the length of their lower leg. As a consequence, they will feel like they are having to prevent the laptop from sliding forward or they have to raise their feet onto their tiptoes to make their thighs level. There are many gadgets that are supplied by the IT market to support the laptop on the lap. However, many of them are bulky and negate the mobility of the laptop itself by becoming a heavy add-on to carry around. Users should think of alternatives. For instance, an inflatable pillow used for camping is about the same size as a standard laptop. It can be blown up when needed and deflated and placed in the same bag as the laptop, adding minimal additional weight to the bag. The only issue the user has to come to grips with is whether they can cope with any embarrassment over blowing the pillow up in public waiting areas or conferences.

If users are required to sit for lengthy meetings in a meeting room, they should consider wheeling their own adjustable chair to the meeting rather than sitting on the non-adjustable conference style seating.

Hotel rooms have items around that can help to increase comfort levels. For instance, sitting on a pillow or folded blanket if the chair is too low will enable the laptop to be used more easily once it is located on a dressing table. Placing a pillow behind the back, if there is only an armchair in the room, will offer support to the user rather than them perching on the front of the chair without support whilst working on the laptop. Rather than make do with a poor working posture, the user needs to scout round the room to see what they can use to make the situation easier for themselves. However, to do this successfully, they need to know what they are trying to achieve, and this information can be imparted during training. Alternatively, many hotels, specifically used by business travellers, have work zones with work surfaces, chairs, and power points. The user should ask when checking in if such a facility is available at the hotel.

5.2.3 Mobile Offices

Many individuals find themselves operating portable equipment in some form of transport: a plane, train, or car. There is little that can be done to make a user more comfortable when in a plane, apart from suggesting that they try to gain as much leg room as possible and that they consider using a small cushion at the small of the back. Obviously, the more a passenger pays for their ticket, the more space and flexibility they will have to make themselves comfortable when working. In fact, airlines make concerted efforts to accommodate passengers who want to work in business and first class. Using a laptop can be a challenge in economy class, especially when the passenger in front decides to recline their seat backwards so they can sleep. At least waiting areas at airports usually provide work zones with surfaces and seating, along with power points and phone charging points nowadays.

The same sorts of restrictions apply when on a train. As little can be done from a set-up point of view, the only option available to control the possible negative impact of poor working conditions is to limit the length of exposure. On that basis, mobile computer users should ensure that they work for short periods only before taking a complete break, which should include stretching or leaving the seated position. If possible, when train tickets are being booked, a specific request for a table should be made so that the mobile device does not have to be placed on the lap or suspended in front of the body without support. An aisle seat will allow for more freedom of movement, especially for the arms, and will lessen the impact of glare from the windows. Travelling first or business class on a train will provide greater space. Prior to boarding the train, access to a premium waiting area is likely to provide access to a working zone.

Using portable equipment in cars introduces other possibilities, as well as other problems. Sang et al. (2010) have indicated that the use of a car as a mobile office may be problematic. Business drivers, therefore, represent a group which may benefit from specific efforts to reduce their risk of musculoskeletal disorders. Many travelling sales reps or consultants carry out preparatory work in their cars having arrived at a client's or customer's site. Many of them finalise documentation or accounts while sitting in the car after leaving the site. Because of the severe restrictions imposed by working within the confines of the car, every effort should be made to avoid the need to do so. Business drivers, for example, who drive for 20 hours or more per week, are considered to be particularly susceptible and are more likely to take sick leave (Porter and Gyi 2002). Reasons suggested for this include constrained postures and seat vibration (Basri and Griffin 2012). Although vibration has no connection with the use of mobile technology, exposure to vibration over time is likely to compound the effects of frequently adopting poor working postures when working in a car.

One way of dealing with space restrictions is to provide users with vehicles that have large interiors, such as vans, that allow for the rear seats to be rotated into a face-to-face position and for stowaway tables to be deployed. However, given the expense of such vehicles when compared with a standard car, this is unlikely to be considered a viable option for many. Clearly, the reality is that some people will continue to, or be expected to, work when seated in a standard car. On that basis, they should be offered information on the likely consequences of working within the vehicle and advice on the most appropriate techniques to employ within that environment. Before the user tries to make themselves comfortable in their car, they should consider whether they could do the same work in a nearby restaurant or coffee shop, where they could use a table.

One of the most immediate concerns with working in a car is security. It is not unknown for portable equipment to be snatched from individuals working in their vehicles. One of the conflicts in this situation is that users of portable equipment should not sit in the driver's seat when operating their equipment, but should sit in the front passenger seat. This is because should another vehicle collide with theirs when they are stationary, it is possible that the airbag might deploy, rapidly and abruptly propelling the equipment towards the user. The steering wheel itself might be moved during an impact, again, causing problems in the minimal space between it and the seated user. On that basis, it is safer to sit in the front passenger seat. However, if the user becomes aware that they are at risk at any point while parked, or simply feels threatened in any way, they cannot start the car immediately and drive away as they could do if they were sitting in the driver's seat. This suggests that drivers need to select a parking area with care; this issue should be tackled during a dedicated drivers' health and safety course that covers all topics of interest to individuals who drive as a significant part of their work.

Before sitting in the front passenger seat, the user should turn off the airbag stowed in the dashboard. The user should then retrieve the laptop from its point of storage, which, ideally, is in the trunk. From a security point of view, this avoids making it a target for theft, and, from a safety perspective, this prevents it becoming an unintended missile following a high-speed impact whilst driving, which itself could cause serious injury. If a laptop is stored in one of the footwells of the vehicle, the user should leave the driver's seat, access the door nearest the footwell, and remove the laptop. Laptops should not be pulled from the rear footwells in the vehicle by someone seated in the front, because of the irregular postures that are likely to be employed when manually handling the laptop and the difficulty in manoeuvring the laptop through the gap between the front seats.

Once seated in the front passenger seat, the user should push the seat backwards to gain maximum legroom. This should also provide sufficient room to place the laptop on the thighs without it coming in contact with the dashboard.

The angle of the backrest should be altered so that the user is supported in the position of choice. If need be, a small rolled-up towel, cushion, or inflatable pillow can be placed at the small of the back to take the place of the lumbar support normally found on a standard office chair. Given the fact that working in a vehicle is far from ideal, the period of time spent working in this environment should be strictly regulated and limited.

There are brackets that can be used to support a laptop inside a car. However, they tend to position the laptop to the side of the user, which causes them to lean and twist to one side. This would not be acceptable for long-term use. This is also the case if the user places the laptop on the passenger seat beside them. The resulting twisted and bent posture would be highly likely to result in backache for the user.

Some businesses have employed touchscreens that can fold down from the interior roof of the vehicle. The positioning of the screen can cause the user to work with at least one arm raised to above shoulder height and to raise their head as they view the screen. If they worked with the screen for extended periods of time, they would be likely to complain of neck and shoulder ache. Sometimes, a separate keyboard is used to input lengthy strings of information and interrogate the system. This screen location would only be suitable for small, intermittent bursts of inputting and interrogation.

Other businesses, such as police forces and utilities companies, have located touchscreens in the dashboard/fascia. The issues of concern on this occasion would be the reaching distances and the readability of the screen in different levels of lighting. In the UK, the location of the screen on the dashboard to the left side of the steering wheel would result in the majority of users having to use their non-dominant hand to touch the screen, or they would have to reach across to their left side with their right arm. The application of sound anthropometric principles at the design stage should help eliminate any problems or limit the impact resulting from poor positioning of the screen due to height and distance.

Another important issue in relation to using cars for work purposes is manual handling. This topic is covered in Chapter 10, which provides specific advice on moving loads in and out of vehicles.

5.3 Summary

- Wireless technology means that people are able to work almost anywhere. Within a building, an individual can work at any point and alongside any other individual.
- The freedom and flexibility offered by wireless technology should be monitored carefully so that individuals do not work in areas that were not originally designed for keyboard work.
- A number of people work away from offices, and whether they choose to work in an office, at home, or in their vehicles, they, too, need advice on the most appropriate way to work.
- People have basically the same requirements irrespective of the environment in which they are working.
- People working in home offices should be provided with appropriate workstation furniture, should the length of time they spend working warrant it.
- Home workers should be offered the same guiding principles that are offered to workers in the main office.
- If home workers do not warrant the provision of specific workstation furniture, they should be given advice on how to adapt the furniture they use at home to make themselves more comfortable while working.

- Users have to exert some control over a transient environment in which they find themselves when working away from the office and away from home.
- Users will need to look for improvised methods to make themselves comfortable when working.
- Users should check if the environment they have arrived at has a specific guest-only work zone, which will provide a more comfortable working arrangement.
- If users are required to sit for lengthy meetings in a meeting room, they should consider wheeling their own adjustable chair to the meeting.
- Working within any form of transport should be limited because of the severe restrictions imposed by the confined space.
- Upgrading seats, or paying for seats with more legroom, when flying may result in more space and a better working environment.
- Work zones in airport waiting areas should be used in preference to working off the lap, if space is available.
- Employees should try to book train tickets that include a seat alongside a table so that their mobile devices can be supported.
- Sitting in a seat adjacent to the aisle provides more room to move the arms and reduces the likelihood of glare being an issue from the window.
- The security issues associated with using a laptop in a car cannot be overlooked and are a primary concern.
- Instead of working in the car, an employee could drive to a nearby restaurant or coffee shop to work so that they can use a table and work more securely.
- Laptop users should sit in the passenger seat of the vehicle, having turned off the airbag.
- It should be understood that laptop users will be unable to leave an area quickly if they feel threatened if they are sitting in the passenger seat of a vehicle.
- Once in the passenger seat, the laptop user should push the seat back to provide sufficient leg clearance.
- The backrest of the passenger seat should be altered to provide adequate support for the user operating the laptop.
- Laptop users should not drag the laptop from one side of the vehicle to another. They should exit through one door and access the laptop through the nearest door to reduce reaching.

- Carrying out laptop work in a vehicle is far from ideal, and the time spent working in this environment should be strictly regulated and limited.
- The application of sound anthropometric principles at the design stage should help eliminate any problems or limit the impact resulting from poor positioning of a touchscreen due to height and distance.

References

Asundi, K., Odell, D., Luce, A., and Dennerlein, J. T. 2012. Changes in posture through the use of simple inclines with notebook computers placed on a standard desk. *Applied Ergonomics* 43(2), 400–407.

Basri, B., and Griffin, M. J. 2012. Equivalent comfort contours for ver-tical seat vibration: Effect of vibration magnitude and backrest inclination. *Ergonomics* 55(8), 909–922.

Heasman, T., Brooks, A., and Stewart, T. 2000. Health and safety of portable display screen equipment. Contract research report 304.

Porter, J. M., and Gyi, D. E. 2002. The prevalence of musculoskeletal troubles among car drivers. *Occupational Medicine* 52(1), 4–12.

Sang, K. J. C., Gyi, D. E., and Haslam, C. O. 2010. Musculoskeletal symptoms in pharmaceutical sales representatives. *Occupational Medicine* 60(2), 108–114.

Sang, K. J. C., Gyi, D. E., and Haslam, C. O. 2010. Stakeholder perspectives on managing the occupational health of UK business drivers: A qualitative approach. *Applied Ergonomics* 42(3), 419–425.

Werth, A., and Babski-Reeves, K. 2014. Effects of portable computing devices on posture, muscle activation levels and efficiency. *Applied Ergonomics* 45(6), 1603–1609.

6

Input Devices

6.1 Introduction

New technology has provided a lot of useful features, in the form of alternative input devices, which were not previously available. However, Noah et al. (2017) have stressed that it is important to recognise that the constant arrival of new technology not only provides commerce and industry with solutions, it also presents new and alternative 'challenges'.

Users are driving the market place, to an extent, in terms of what they are looking for from a device and the facilities it offers. Users prefer, and expect, high input speed (Sandnes and Aubert, 2007) and ease of learning when presented with new technologies. They are not averse to showing a reluctance to buy into products that are perceived as overly cognitively demanding.

Users no longer have to rely on using external input systems, such as keyboards or mice, to interact with their computer system. They can use internal input devices, such as virtual keyboards on their tablets, trackpoints (the small joystick embedded in the keys of some laptops), and touchpads (which is a motion-sensitive pad normally located at the front and centre of laptops).

A differentiation also has to be made between direct and indirect input devices. Direct input devices, such as a touchscreen, permit direct control over the process being carried out. With indirect input devices, such as a touch pad, the process becomes cognitively more demanding. Indirect input devices require users to map the cursor movement on the screen, which results in a more complex and specific transformation. This is partly due to the fact that the user's on-screen displayed actions will be remote from the point where the changes are instigated. With a touchscreen, the instigation and action occur at the same point. It has been suggested (Schürmann et al. 2015) that device developers do not currently seem to separate the two input methods and, instead, they seem to treat the operational concepts as equal.

It would appear that ageing is an important factor to be considered when developing or choosing an input device design. Age-related changes in sensory, as well as motor, performance is well known. Older adults seem to demonstrate in some studies (e.g. Armbruster et al. 2007) a better performance when using a direct input device, like a light pen, as it allows an easy mapping operation. Jochems et al. (2013) would state quite specifically that they would recommend a touchscreen for the elderly. However, distinctions do need to be made between those considered to be 'elderly' and those considered to be part of an 'older' generation. Elderly people may only have started to get to grips with basic computing systems in their 70s or later, whereas anyone under 70 years of age is likely to have used computers as a significant part of their working life, even if they did not use them at school. Because direct input devices are simpler to use than indirect devices in terms of the psychomotor skills required, this has implications in the case of selecting inputting methods for use by disabled users.

Even within one single item, such as a mouse, there are many options available with regard to design and the technology used to operate the device. As is the case with every other item used in an office setting, these pieces of equipment have to be selected with care, following thorough consideration of the implications of their design features and how they are used. For instance, non-keyboard input devices (NKID) are generally operated in fairly static postures, and the movements required are concentrated on the fingers, wrist, arm, and shoulder, but, it is apparent from a small number of studies that different devices require different postures to operate them (Woods et al. 2003). Appropriate selection and trialling of options are advisable, even for such seemingly simple equipment, which many people view as of secondary importance. It should be kept in mind that these pieces of equipment will have a fundamental impact on the postures adopted by users when working—not only upper limb posture, but whole body posture as well—and workers will probably be interacting with them consistently over an eight-hour period or longer.

There are a number of tasks that users perform with input devices, and each device varies in how well it performs these tasks. Therefore, before selecting a specific device, consideration has to be given to the task demands. Questions need to be asked, such as whether the device will be used to simply point at an object on the screen or select it for further action, or perhaps track the movement of something presented on-screen, or whether items will be dragged from one area of the screen to another, or whether the user will position or reorientate objects, or whether they will input or edit information.

The basic factors considered important for good design of non-keyboard input devices (NKID) such as mice, trackballs, joystick, etc. are:

1. Comfortable hand and finger position;
2. Adequate control;
3. Intuitive and easy-to-use. Intuitive interaction with a product is defined as the unconscious application of knowledge on the side of the user (Mohs et al. 2006);
4. Ease of device, button, and trackball movement;
5. Good interaction with software; and
6. Provision of suitable accessories.

6.2 Keyboards

The conventional, straight keyboard is still the most popular design sold and used with personal computers, despite what are considered to be the biomechanical benefits offered by alternative designs.

It has been established (Rempela et al. 2007) that wrist and forearm postures are strongly influenced by keyboard design. The flat design of the QWERTY keyboard requires users to make postural adaptations to conform to the keyboard. Users have to work with their hands pronated (with palms facing down) and in ulnar deviation (turned at the wrist in the direction of the little finger) (see Figure 6.1). It has been suggested that depending on the specific model of keyboard in use, this ulnar deviation can be up to 25° from neutral (Zecevic et al. 2000). It has been suggested that the smaller the keyboard, the more irregular the posture (Pheasant and Haslegrave 2006). This is significant with regard to laptop computers and notebooks. Irregular upper limb postures have been associated with the development of upper limb disorders. This issue will be dealt with in more detail in Chapter 11.

Previous studies (Kim et al. 2014) have shown that a keyboard's key activation force, travel distance, force-displacement characteristics, and tactile feedback can affect typing forces, muscle activity, muscle fatigue, and discomfort in the upper extremities. Typing forces increase with higher key activation forces, and higher typing forces result in increased muscle activity, muscle fatigue, and discomfort in the upper extremities.

Due to the fact that tablet use is much more common, users are working with virtual, touchscreen keyboards with no physical key feedback. This

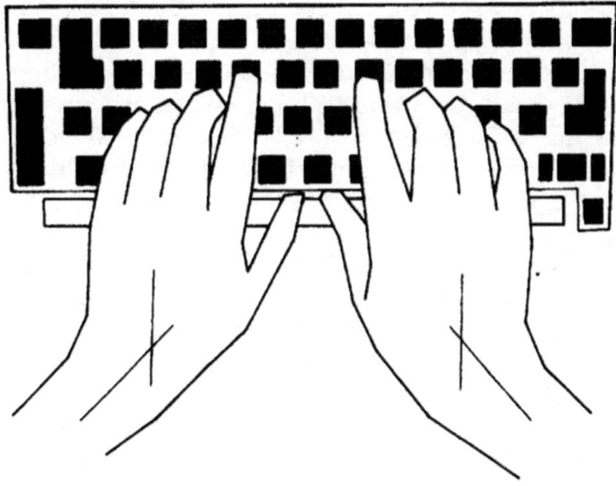

FIGURE 6.1
The posture of the hands dictated by the design of the keyboard. (Bridger 2017).

may have an impact on typing productivity and the physical risk factors associated with upper limb disorders (ULDs). A touchscreen keyboard is completely different from conventional separate keyboards in terms of key feedback characteristics. Standard separate keyboards have a narrow range of activation forces, typically between 0.5 and 0.8 N, which means that users can rest their fingers on the keyboard keys. In contrast, keys on a virtual keyboard are activated by any physical contact with the skin, so users are unable to rest their fingers on the keyboard, which means they suspend their hand above the keys to avoid accidental activation. This posture could lead to extended static loading in the finger/forearm extensors and shoulder muscles, and could result in an increase in risk. Kim et al. (2014) are of the view that, because a virtual keyboard provides limited tactile feedback without key travel and force-displacement characteristics, there may be differences in typing productivity and typing forces when compared with standard separate keyboards, which provide users with some sort of tactile feedback. For that reason, they believe that for long typing sessions, or when typing productivity is a priority, conventional keyboards with tactile feedback may be more suitable even though typing on a virtual keyboard is done with less force. This is due to the fact that they found that lower typing forces and finger muscle activity when using a virtual keyboard came at the expense of a 60% reduction in typing productivity.

Although forces required to press computer keyboard keys are low, the repetitive nature of typing may result in enough cumulative fingertip force for those who type extensively throughout the workday to contribute to upper limb disorders (ULDs), such as carpal tunnel syndrome and tenosynovitis. Fingertip force was identified quite some time ago as being associated with the development of upper limb disorders. Bufton et al. (2006) showed that although users applied less fingertip force to notebook keys than to desktop keys, they did apply greater overstrike force (i.e. more force than is necessary to successfully depress the keys) on notebooks than on desktop keyboards. Bufton et al. suggested that overstrike force when using notebooks might be reduced by increasing tactile feedback and travel distance in keys. This has implications for the selection of notebooks, suggesting that consideration should be given to the form and extent of feedback given to users when they are operating the keyboard.

The unnecessary use of greater force to depress keys, which, in effect, unnecessarily increases the workload for the upper limbs, suggests that this is an area that should be tackled by training in appropriate keyboard skills. One of the drawbacks resulting from the common use of computers is that users do not need to have acquired specific typing skills to be able to perform a computer task. As a consequence, many users have a tendency to employ keyboard techniques that are likely to increase their workload and, ultimately, the likelihood of them experiencing discomfort.

Because one character per key is the norm for standard English entry keyboards, which results in a requirement for 36 keys, miniature keyboards shrink the keys to fit them into a limited space. This has been identified as creating difficulties and discomfort for users. As an alternative to shrinking the keys, one study (Hsiao et al. 2014) developed a rectangular-shaped keycap containing 3 letters with separated keycaps of numerals. Results identified that users obtained proficient speed, higher comfort and greater acceptance than many other alternative designs whilst avoiding having to use small individual keys.

When users obtain information from portable devices, they prefer the largest possible display area and the smallest keyboard that provides good usability and typing speed (Pereira et al. 2013). Smartwatches have been developed with similar functions to a smartphone. However, the small size of the device restricts the screen size and the information space, and also it makes it difficult to input text. To get around this problem, a Virtual Sliding QWERTY, which utilises a virtual qwerty-layout keyboard and a 'Tap-N-Drag' method to move the keyboard to the desired position, has been developed (Cha et al. 2015). It has been reported that this requires very little learning time for users and is intuitive to use because it utilises the full sized QWERTY layout. Despite limitations in the size of the screen, it is possible to access any key on the keyboard by sliding the keyboard using the finger in any direction. There is another strategy for utilising a smartwatch called ZoomBoard (Oney et al. 2013), which is based on a QWERTY keyboard layout and a zoomable interface. However, it would appear that users find this time consuming, because they have to zoom in/out repeatedly to select letters.

The QWERTY keys are allocated in a way that results in an unequal distribution of the load between the hands, assuming 'trained' usage of the keyboard. Only 32% of English words are typed using the home row of a standard keyboard. There are high levels of 'row hopping' as users move mainly from the top to the bottom row of keys as they input frequently used sequences of words. This reflects the extent to which the hands have to travel when operating a standard keyboard rather than an 'ergonomically-designed' keyboard, which suggests the upper limbs may be unnecessarily overused when operating a standard keyboard. Zecevic et al. (2000) have stated that changes in keyboard design could not only improve hand position, but would also maintain performance at the same level, or even improve it.

It has been suggested that split keyboards may improve comfort and reduce fatigue. These keyboards aim to present the keys so that they are aligned with the natural resting position of the hands and fingers and with more natural positioning of the arms. This is in contrast to the standard QWERTY keyboard, where users have to adjust their finger, hand, and arm positions to depress the keys. Alternatively, contoured keyboards have been shown to reduce levels of finger and wrist muscle activity.

The provision of keyboards with integral number pads should be reviewed. Integrating a number pad onto the side of the keyboard layout increases the width of the keyboard. As the mouse is generally placed to the side of the keyboard, this layout results in the mouse being further from users than they might wish (see Figure 6.2). This can contribute, over time, to discomfort. Users can experience even greater discomfort if they are also sitting too low relative to their desks while using the mouse. Unless an individual is a regular user of the number pad, it would be beneficial to provide a keyboard without a number pad, provide a keyboard with a pull-out number pad, or provide a separate number pad altogether. This will allow the user to place the mouse closer and reduce the extent to which they have to move their arm out to the side (see Figure 6.3).

It is worthy of comment that many people are not explicitly aware that the number keypad layout differs across devices, such as telephones and calculators. An interesting observation was made in a study by Armand et al. (2014) that their results indicated that participants' memory of the layout for the

FIGURE 6.2
The positioning of the mouse and related arm posture resulting from inclusion of a numberpad in the keyboard.

FIGURE 6.3
The position of the mouse and posture of the arm following removal of the numberpad.

arrangement of keys on a telephone was significantly better than the layout of a calculator. However, the results showed that participants were more accurate when entering stimuli using the calculator keypad layout. This has significance if the user is working in a department where they input numbers extensively.

The position of the keyboard relative to the desk edge is important. Sufficient space should be left in front of the keyboard so that the hands and forearms can be rested in-between bouts of keying. However, the space should not be so great that it causes the user to overreach to depress the keys. A 10 cm space in front of the keyboard is usually enough.

The keyboard can be used to reduce mouse work in situations where a user is experiencing upper limb discomfort associated with excessive mouse use. Individual keys can be assigned as accelerators (also called short cuts or macros) that allow the user to access menus without having to click the mouse to display the menu and select from it. This is likely to be most successful with experienced users. However, it should be kept in mind that this system relies more heavily on memory. A surprisingly small percentage of users know that the function keys will permit them to execute a menu-driven task without using the mouse.

A greater proportion of males exhibit altered shoulder postures than females while using both keyboards and touchscreens on mobile devices. It has been suggested (Gold et al. 2012) that as males have typically larger shoulder breadths than females, they may find it more difficult to access the small keys on these devices without protracting their shoulders.

6.3 Mouse

Mice come in a variety of forms, such as the standard mouse, the vertical mouse, and the trackball mouse. In terms of the psychomotor skills required to operate it, the mouse requires a lower skill base for successful use than other input devices, such as the trackball, because there is a close proportional relationship between how far the mouse is moved, the direction in which it is moved, and the speed with which it is moved and with what occurs on the screen in terms of cursor movement.

One of the major difficulties with mice is that many have not been designed using basic hand-tool design principles, whereby the object being gripped and operated is actually designed to conform to the dimensions, shape, and capabilities of the hand. Many mice are simply a rectangular lump that does little to assist the user in terms of comfort or ease-of-use. Despite the variety of users who are likely to use them, very few mice provide a means to adjust their size to accommodate the hand of the individual. The same sort of criticism can be levelled at a puck, which is similar in shape to a mouse and is used with a digitising tablet.

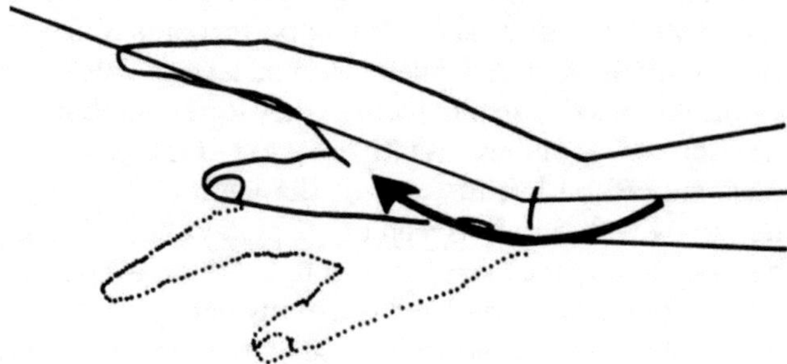

FIGURE 6.4
The movement of the hand away from a neutral position when using a mouse. (Adapted from BS EN ISO 9241-9:2000. Ergonomic requirements for office work with visual display terminals (VDTs). Requirements for non-keyboard input devices.)

The design of the standard mouse generally requires a fully pronated forearm during its operation, so the palm of the hand faces down and the wrist is extended, where the hand is raised upwards at the wrist away from the neutral position (see Figure 6.4). Some mice are so large, either in length or width, that the buttons are difficult to reach, which can result in repeated repositioning of the hand on the mouse or overextension of the fingers to reach the buttons. Some mice are designed to be operated by the right hand, but left-handed users employ them on the left side of the keyboard. The right-handed mouse obviously will not fit the contours of the left hand satisfactorily. A mouse designed so that it can be used with a neutral hand position will significantly reduce musculoskeletal symptoms among computer operators.

A particular problem with the mouse is that its location is usually secondary to the keyboard. In other words, it does not get prime positioning on the desk so that it is close to and in front of the user. It is relegated to the side of the keyboard, and this often causes poor posture in the 'mousing' arm. The actual wrist position is heavily influenced by the location of the mouse relative to the seated user, more than by the design of the mouse itself. Dennerlein and Johnson (2006) found that there are differences in exposure to various biomechanical risk factors that are dependent on where the mouse is positioned—for example, to the side of the user, in front of the user or at a distance from the user. Use of the mouse to the right of a standard keyboard can lead to an increase in muscle tension in the upper shoulder, back, and arm. Irregular wrist and shoulder postures can also be employed for a greater percentage of time when using the mouse than when using the keyboard. Operating a mouse also requires greater static posture of the hand and fingers than is the case when operating the keyboard. The greatest movement occurs when clicking, and there is some side-to-side movement of the wrist as the cursor is repositioned or as an item is dragged. The use of a mouse is considered to be a risk factor for hand and arm disorders (Funch Lassen et al. 2004), and increased prevalence of neck and shoulder disorders in the right side of the body has been associated with mouse use. A study by Woods et al. (2002) showed that 42% of mouse users reported pain or discomfort associated with use of this device. It has also been suggested that four hours of intensive computer work using a mouse results in muscle fatigue in the forearm muscles without signs of recovery after two hours of subsequent rest.

Some users position their mice in even more extreme positions on the desk, because they are working in cramped conditions brought about by the number of items they are working with simultaneously, or by a work surface layout that leaves little 'mousing' space. As a consequence, they move the mouse to the outer extremes of their reach, such as alongside the screen. This is a prime example of how training, or lack of it, can influence the working practices adopted by users. Providing an L-shaped desk may supply the additional space required for unhindered use of the mouse, and this type of desk may also offer some support to the forearm. A 'wraparound', or curved, L-shaped desk is preferable to one that forms a 90° angle where the two lengths of desk meet. If users do work at L-shaped desks, it is important that they sit at the appropriate height, as outlined in Chapter 1. If they sit too low relative to the desk, they will have to raise their arms and probably abduct them (i.e. move the arm out to the side) to rest their forearms on the desk surface as they use the mouse. This is likely to result in discomfort in the 'mousing' arm.

Many users work for years without any knowledge of the simple steps that maintain comfort levels. For instance, users need to be advised to take their hands off the mouse when it is not in use. This will allow them to return their hands to a more neutral resting position. Some users may wish to use a mouse mat with an integral wrist rest. Caution should be observed when this type of mat is being used. On the one hand, the integral wrist rest supports the wrist and ensures that the hand and wrist remain in a neutral position. However, once a user leans on this wrist rest, their arm is, in effect, anchored in place. This will mean that the mouse will be more likely to be repositioned by side-to-side movement of the wrist rather than by larger movements of the arm, with the latter being less fatiguing. Users of this type of mat should be monitored. As an alternative, users of optical mice that do not require mouse mats could try using a more mobile type of wrist rest, such as those that are styled like an ice hockey puck. They have a cushioned surface to support the wrist, and they glide across the desk surface as the mouse is moved and repositioned. It works effectively, because it combines support of the wrist with enabling the whole arm to be moved. Gliding palm supports, that are designed to match the contours of the heel of the hand, work in a similar manner.

The mouse is used to achieve a number of different objectives. It is considered to be most effective at pointing, selecting, drawing, and dragging. During dragging, the button is depressed for a greater percentage of the working time than with the other activities, and the pinch forces on the side of the mouse are about three times greater. There is greater fingertip loading of longer duration during a dragging task than during pointing and clicking, which would suggest that it would be advisable to reduce the extent of dragging involved in mouse use.

An alternative form of mouse is the roller bar mouse. Muscular activity was found to be significantly lower for the roller bar mouse than for the conventional mouse (Kumar 2008). A roller bar mouse allows the user to work closer to the body compared to the conventional one, thus, the former can be recommended as a general means of reducing upper extremity musculoskeletal disorders. The degree of tension needed to click the mouse can also be reduced or increased to suit individual users. Lin et al. (2015) found that a roller bar mouse allowed for a more neutral hand posture with greater finger flexion and smaller finger spread compared to using both a mouse and a trackball. In addition, a more neutral hand posture with smaller index-middle finger spread and greater middle and ring finger flexion was associated with the roller bar mouse when compared to a touchpad. This may be explained by the design of the roller bar mouse, which allows multiple fingers to operate the device.

The current design-driven problems associated with standard mouse use suggest that they should not be used for prolonged, uninterrupted periods. Van Niekerk et al. (2015) suggest that users should be advised to use a range of different input devices to encourage postural dynamism whilst sitting. Postural dynamism refers to the number of frequent involuntary postural or movement changes whilst sitting. This is because they established that postural dynamism of the thoracic and cervical spine when sitting is less during mouse use compared with keyboard use.

6.4 Touchpads

Most often, integrated touchpads are located in the computer's wrist rest and centred on the keyboard. The touchpad has to be stroked by finger movements which are translated into a matching cursor movement. The high correspondence between hand and cursor movement make the control of the cursor of the touchpad an easier task, compared to other non-keyboard input devices, such as, for example, a trackpoint (Armbruster et al. 2007; Sutter and Oehl 2010). The trackpoint senses force from the fingertip, which results in a cursor movement which may not mirror the directional movement of the finger operating the point. It is believed that the ease with which the transformation of hand movement into cursor movement can occur will determine the efficiency of user-computer interaction (Sutter and Ziefle 2006). This has implications for business in terms of what methods of inputting device are not only provided by, but are also promoted by the business. Although both the trackpoint and touchpad are operated very differently, they still have to meet user demands for speed and accuracy equally well (Wu et al. 2013).

It has been reported (Kelaher et al. 2001) that touchpad location has a significant impact on upper extremity posture, discomfort, preference, and performance. Studies suggest that the most common location for touchpads in laptops may not be the optimal location for it. The bottom-centre, bottom-right, and right-side locations get the best ratings from users, although subjects' wrists were more extended in bottom locations more than other positions. The right-sided preference for location of the touchpad might support the use of a clip-on attachment, an external device, or even integration into the side of the laptop.

Because the positioning of the touchpad can result in users adopting an upper limb posture that deviates from the neutral, lengthy use of the touchpad should be avoided by encouraging users to operate a separate mouse, in the alternative.

6.5 Trackball

Trackballs have been referred to as 'upside-down mice'. Because they are used in this orientation, they do not move around on the desk, so their space requirement is not as great as a standard mouse. A trackball rotates freely within its housing in all directions. Depending on how the trackball is presented, it can be activated by the thumb, fingers, or palm. As the ball is rotated, its direction and speed of movement are tracked, and this is translated into a proportional movement of the cursor on the screen. It is considered to be most effective when used for pointing, selecting, and tracking.

Working with a trackball that is centred at the bottom of the keyboard, such as on a laptop, results in lower levels of activity in the shoulder muscles than is the case with mouse work out to the side of the keyboard. However, it results in the most extreme wrist extension when compared with other input devices. This is due, in part, to the fact that a number of trackballs are provided with built-in support for the hand, but, as the support is elevated above the desk, this causes the unsupported arm to 'droop' onto the work surface. The degree of extension will obviously be related to the specific design features of the model in use. The trackball is sensitive to touch, so users tend to work with their fingers suspended above the trackball when not using it to avoid inadvertent activation. This is likely to be fatiguing. Rolling forces on the trackball of between 0.2 and 1.5 N are usually considered acceptable.

Some trackballs are 'embedded' in standard mouse shapes and rely on the ball being activated by the thumb alone. Reliance on the thumb to execute rapid repetitive movements, when it is held away from the main body of the hand, should be considered carefully. Combined actions and postures such as these could lead to discomfort for the user.

6.6 Vertical Mouse or Joystick

Vertical mice or joysticks are stick-shaped and are anchored at the base. A larger joystick operated by the whole hand is intended to allow the user to adopt a more neutral position of the forearm. It is designed so that the user can employ a power grasp rather than a precision grip. However, using such a grasp may have an impact on the precision with which it can be operated, since using the whole hand in this manner generates more extreme movements. A joystick is most effective when used for selecting and tracking. Therefore, it will probably work effectively in only a few applications.

Some designs rely heavily on the thumb to operate buttons at the top of the stick, and extended use of the thumb in this manner is not advisable. Smaller joysticks are operated by the fingers alone and respond to fine movements requiring greater motor control.

The direction and distance over which the joystick moves is mirrored in a corresponding movement of the cursor on the screen. If the joystick is used in conjunction with a keyboard, the user has to learn to switch between two completely different ways of using the hand, which can slow them down. It is generally considered that the displacement in a hand-held joystick should not exceed 45° to the left and right, 30° away from the user (forwards), and 15° towards the user (backwards).

6.7 Touchscreens

A differentiation has to be made between touchscreens, such as fixed vertical screens that might be used on a desk, as part of a cash dispenser, or line-side screens used in a production/assembly environment, and touchscreens used on mobile devices such as smartphones, phablets, and tablets. The circumstances in which they are used are very different. Orphanides and Nam (2017) believe that the development of touchscreens should be done within the system context, i.e. making reference to who is using them, where they are using them, and what they are trying to achieve by using them.

Touchscreen interfaces are known to have advantages over traditional pointing devices in interacting with the computer by enabling users to interact directly with an item, or multiple items, that are displayed without any intermediate devices such as a mouse (Forlines et al. 2007).

Unlike interaction with devices that require only finger and wrist movements, such as mice and trackpoints, interacting with fixed vertical touchscreens can require whole arm movements, often without hand or arm support. Users may not be able to maintain neutral arm and hand postures, or rest their arms on chair armrests or a table top if the touchscreen is positioned vertically. The use of a touchscreen is often associated with a significant increase in subjective discomfort in the shoulder, neck, and fingers, increased activity of shoulder and neck muscles, and a greater percentage of time with the arms in the air. Common guidelines for computer workstations, such as placing the keyboard at elbow height and having the top of the viewable area of the screen at, or slightly below, eye-level, do not work for fixed vertical touchscreens. If the touchscreen is placed horizontally or on a slant at a lower level, it may cause greater neck discomfort than when it is positioned more vertically at higher locations (Shin and Zhu 2011). If vertically presented touchscreens are to be used from a standing position, they should be positioned at a height where the user does not have to reach above shoulder level. Horizontally presented touchscreens should be no higher than elbow height.

Proper target location (i.e. the specific spot that needs to be touched to make a selection) and display positioning on the touchscreen are important as they influence task performance and physical demands of fixed vertical touchscreen use. Placing frequently accessed targets in the lower area and visual content in the upper area within the display will reduce both shoulder and neck muscle fatigue while achieving an acceptable level of touch accuracy during repetitive tap gestures (Kang and Shin 2017). The size of an individual item of information displayed on the screen needs to be considered carefully. It needs to be of a size that can be activated easily by users with either small or large fingers. It also needs to be placed so that inadvertent activation is unlikely, and the layout needs to be arranged so that priority information is not located at the bottom of the screen where it might become obscured by the outstretched hand. As the hand comes in contact with the screen on a regular basis, this produces a requirement for more frequent cleaning. If users employ an object, such as a pen, to touch the screen, this may cause damage. It is likely that auditory feedback as the screen areas are touched will facilitate performance.

Positioning the display upright and close to the user would help them complete the tap gesture faster with less muscle activity, but it is likely that more frequent rest breaks would be needed to avoid eyestrain from the short viewing distance. This suggests that their use is not advisable for prolonged inputting or interaction with the screen. They might be better suited to short selection-type tasks in which an individual selects from a menu and where there is little or no text inputting.

When assessments are being made of users' interaction with the touchscreens of various mobile devices, it should be kept in mind that tapping may not be representative of the diverse gestures necessary for multi-touch interaction. Smart phones, tablets, and in-car navigation systems include touchscreens that require multi-touch finger taps, slides, and other gestures (Asakawa et al. 2017). Multi-touch gestures may involve greater joint excursions than tapping, potentially increasing the use of extended or flexed thumb and wrist joint postures, which are associated with decreased performance (Trudeau et al. 2012).

It is thought (Berolo 2011) that touchscreen technology could potentially contribute to repetitive strain injuries or other musculoskeletal disorders related to the overuse of the fingers and thumb. Flicking and panning gestures used for in-vehicle navigation systems require more time than tapping, potentially increasing musculoskeletal exposure to loads (Kim and Song, 2014). It has been shown (Dennerlein, 2015) that software interfaces that require fewer pinch and stretch gestures, or smaller pinch and stretch gestures, may potentially reduce strain on the index finger and thumb joints. Gestures, such as taps and slides, may reduce musculoskeletal strain, as the joint postures of the fingers are more neutral, avoiding excessive flexion or extension of the finger joints.

6.8 Graphics Tablets

When graphics tablets and pens are used, there is a direct relationship between the direction, speed, and distance of movement on the tablet and the cursor on the screen. However, as the user has to move between two different ways of using the hand if swapping between keyboard and tablet, it may take some time to develop an appropriate skill base. The tablet and pen are most effective when used for pointing, selecting, dragging, and drawing. The contact surface of the tablet should be flat and smooth and devoid of reflection or glare.

A pinch grip is employed to hold the pen, and extended use of this particular grip is likely to be fatiguing, so extended uninterrupted use should be avoided.

The recommended diameter for a touch-pen when performing pointing- and clicking, writing, and drawing tasks is 8 mm. A length of 100 mm or more is suitable for all pen-based devices (Wu and Luo 2006).

6.9 Voice Recognition Software

Voice recognition software (VRS) is an alternative tool to using more traditional input device, such as keyboards and mice. Many users who have disabilities, or who have developed upper limb disorders, turn to VRS as a means of continuing in their employment without the need to rely solely on their hands to enter or retrieve data from the system. Users with dyslexia have also used this system in preference to keyboards. It is worthy of note that dyslexics can use text-to-speech systems that read back the text that has been inputted, which avoids the need for the dyslexic user to proofread the text displayed on the screen.

For VRS to work effectively, it has to be accurate and work at a speed appropriate to the work rate required of the user. Introducing such a system is not straightforward and requires planning and effort on the part of the user, because the system has to 'learn' to interpret what the user is saying as they speak, given their pattern of speech, its pitch, rate, and inflection. Continuous speech software is the preferred option for users rather than discrete speech recognition software, which is slower and not as natural. Some organisations opt for the latter type because it is cheaper. It is possible to select a package that includes specialised vocabularies. Some systems will also respond to voice commands for controlling the computer, such as running particular programmes and working with particular menus.

Before starting to work with the VRS system, the user has to train the system to recognise their voice. This is achieved by reading aloud for about 30 minutes into a microphone during the 'enrolment' phase. Some specific VRS systems require up to 75 minutes of reading time. The software analyses the sounds and converts them to text. The reading material is preselected and is normally displayed on the screen. Once the user has completed the enrolment phase, they can start to use the system.

To utilise the system effectively, users must learn to talk fluently and clearly so that the VRS can recognise what they are saying. This is more difficult than might be imagined, as during normal speech an individual will often mumble, hesitate, or slur words. Once this system is in use, users should gather their thoughts before speaking and then talk in complete sentences. Proofreading will be required, since the system will misinterpret words, and the errors will have to be corrected. Some VRS users are under the impression that although they might input the main body of their text by voice, they may have to edit material by using the keyboard and mouse. This is not the case, as the system has been set up so that editing can be achieved using the voice. The system should learn from the correction process, which will improve future accuracy.

From a practical point of view, consideration has to be given to the environment in which the system will be operated. Although microphones are adept at filtering out background noises that might interfere with speech interpretation, very noisy backgrounds, such as in a call centre, may require a higher-specification microphone, or the individual may need to be moved to a quiet area. This might also be needed if the user deals with confidential material or, alternatively, if other users find VRS use distracting. It should be kept in mind that if users are unwell, for instance with a cold, this may change their voices to the extent that misinterpretations by the system occur.

As might be expected, it has been established that the hand is less tied to the keyboard and mouse during the use of VRS (Juul-Kristensen et al. 2004). However, it has also been shown that the eyes and head do remain in a fixed position for a longer period of time as a result of viewing the computer screen. Reports of overuse of the voice are also becoming more common. This can be combatted by ensuring that, as is the case with all repetitive work, users take regular breaks and perform a variety of tasks, and they should also speak in a normal voice, rather than a raised voice when using the system. As a consequence of the possible adverse effects resulting from overuse of VRS, it has been suggested that VRS should be used more as a supplementary tool to traditional computer input devices as a means to reduce the risk of overloading the muscles in the forearm, shoulder, and neck (Juul-Kristensen et al. 2004), rather than being the sole means of inputting or retrieving data.

6.10 General Design Requirements

Whatever input device is selected, those that are operated by the hand should suit the shape and contours of the hand and fingers. The device should move smoothly, be easy to control and operate, and not require excessive effort. It should allow for precise manipulation, respond promptly, and be capable of being used intuitively. The settings should be capable of being changed without difficulty. To ensure that these requirements are met, it is recommended that any buttons incorporated into the main body of the device be designed so that users can avoid inadvertent activation when their hands or fingers are in the 'at rest' position. However, this should not result in excess force being required to depress the buttons. Displacement forces of between 0.5 and 1.5 N have been suggested. The buttons should be shaped and positioned so that they align with the fingers and can be depressed easily. A displacement of about 5 mm is likely to provide suitable feedback to the user, this feedback being 'experienced' through the fingers. If a user has to hold a button down for extended periods to complete a task, such as drawing, consideration should be given to providing either a software or hardware feature for locking it in the desired position. When buttons are being depressed, this should not result in inadvertent movement of the device.

Left- and right-handed versions of the input devices should be available to users. Some users may choose to alternate the hands used for operating an input device, and the design of the device may necessitate the provision of two separate units. Care should be taken if users start to use their non-preferred hand for long periods, in case this provokes discomfort in this limb. Although most users are able to build up an appropriate skill base with their non-preferred hand, this limb tends to have a lower workload, generally, and may not have developed an appropriate level of 'work hardening' to meet the demands of the task without encountering difficulties.

The area of the device that comes in contact with the hand should not present angled or sharp edges to the user to avoid pressure points. The material used should be shaped and textured to minimise the possibility of the hand slipping. The weight of the device should not affect the ease with which the task is completed.

If the device is to be cabled, as opposed to wireless, consideration should be given to the length of the cable so that it does not limit the user's choice over where it is positioned. The weight and tension of the cable may influence the perceived directional 'pull' of the device when in use, and this type of interference should be eliminated. Wireless input devices offer the greatest level of flexibility over where they are located.

Response time between activation of the input device and reaction on the screen should be such that the user does not perceive a delay. A signal speed of about 20 ms should have no impact on performance. Feedback should be immediate, perceptible, and understood by the user. It should also be consistent.

The device should be presented so that it can be gripped, manoeuvred, and operated quickly and easily without any degradation in performance. This will be influenced by the design of the input device itself, the design and layout of the workstation, and the position of the user relative to the device.

Given that there are so many options among input devices, it is advisable to trial them before introducing them on a large scale. As was suggested in Chapter 3 in the section relating to trials of workstation furniture, feedback forms should be used during the trial phase. One form is completed after every device is used. The device should be used during normal working procedures for about four hours, with appropriate breaks during that period. Care should be taken that users included in the trial adopt a suitable working posture during the trial so that there is no bias from a source removed from the input device itself.

6.11 Summary

- Input devices need to be selected with care following a thorough consideration of their design features.
- Input devices should be trialled before being introduced for everyday work.
- The use of input devices has an impact on the upper limb and whole body postures adopted by the user.
- A number of postural problems are related to the design and layout of standard QWERTY keyboards.

- The use of a QWERTY keyboard causes the user to work with the palm facing down and the hand turned at the wrist in the direction of the little finger, which is a rather stressful posture.
- Smaller keyboards tend to result in more irregular limb postures. This has significance in terms of laptops and tablets.
- The QWERTY keyboard results in high levels of row hopping as the user moves from the top to the bottom row of the keys. This is an inefficient means of working.
- Split keyboards aim to present the keys so they are aligned with the natural resting position of the hands and fingers.
- Contoured keyboards have been shown to reduce the levels of wrist muscle activity.
- Slimline keyboards are considered to be more likely to result in more suitable hand and wrist postures.
- Fingertip force is associated with the development of upper limb disorders.
- Keyswitch design has been shown to affect fingertip force applied to keyboards during key depression.
- Keyboard users may type with more force than is necessary. This is referred to as overstrike force.
- Providing suitable auditory and tactile feedback may reduce overstrike force.
- Laptops and notebooks tend not to provide the same level of auditory and tactile feedback as standard QWERTY keyboards. This may explain why the greatest overstrike forces occur with these types of keyboards, rather than with QWERTY keyboards.
- Training should be used to address the problem of overstrike force.
- The inclusion of a number pad on the keyboard results in irregular positioning of the arm when the mouse is in use. Unless the number pad is in use frequently, a keyboard without one should be provided.
- Keyboards should be positioned approximately 10 cm away from the leading edge of the desk to provide sufficient space to rest the hands and the forearms between bouts of keying.
- The keyboard can be used to reduce the amount of mouse work performed. Individual keys can be assigned as accelerators (shortcuts or macros) that allow the user to access menus without having to use the mouse. This system relies more heavily on memory.
- The design of the standard mouse requires a fully pronated forearm, with the palm facing down, and extension of the wrist, with the hand raised up at the wrist.

- Left-handed users are frequently forced to operate right-handed mice. If the mouse is contoured, it will not fit the contours of the left hand and is not a satisfactory arrangement.
- The mouse is usually given a position secondary to the keyboard. As a consequence, the mousing arm regularly ends up in a poor posture.
- Operating a mouse requires greater static posture of the hand and fingers than is the case when operating the keyboard.
- Users need advice on how to keep their mice clean and how to alter the speed setting using the software facilities. Many users are unaware they can change these features.
- Touchpads are common on laptop computers and are used in place of a standard mouse.
- The positioning of the touchpad can result in irregular postures of the hand and the arm.
- Trackballs are considered to be upside-down mice.
- A trackball is activated by the thumb, fingers, or palm. It results in the most extreme wrist extension when compared with other input devices.
- As a trackball is sensitive to touch, users tend to work with their fingers held above the trackball when not using it, which is fatiguing.
- Heavy reliance on the thumb to operate small trackballs should be reviewed carefully, owing to the association with overuse of the thumb.
- The use of vertical mice or joysticks allow users to adopt a neutral position of the hand and forearm. They are also able to employ a power grasp rather than a precision grasp.
- Employing a power grasp on a vertical mouse or joystick results in more gross movements and may limit its application in situations where fine movements are required.
- Smaller joysticks can be operated by the fingertips and will enable more fine movements to be made.
- Touchscreens are one of the fastest means for pointing at stationary objects on a screen.
- One difficulty in using a fixed vertical touchscreen is that it results in users working with their arms outstretched and unsupported. This suggests that touchscreens are not suitable for extended uninterrupted use.
- The size of individual items displayed on a touchscreen should be considered carefully. They should be arranged so there is no

inadvertent activation. As the hand comes in contact with one part of the screen, it should not obscure information presented in another part.

- Auditory feedback as the touchscreen is used should facilitate performance.
- Vertically presented touchscreens operated from the standing position should not be presented above shoulder level.
- Horizontally presented touchscreens should be no higher than elbow height.
- The use of a graphics tablet and pen requires a pinch grip to hold the pen. This could be fatiguing during extended uninterrupted use. Users should be aware of the need to put the pen down at regular intervals.
- For voice recognition software (VRS) to work effectively, it has to be accurate and work at a speed appropriate to the work rate required of the user.
- The VRS has to learn to interpret what users are saying as they speak, given individual patterns of speech, pitch, rate, and inflection.
- Before starting work with VRS, a user has to train the system to recognise their voice. This involves reading preselected material.
- During the enrolment phase, the user must speak fluently and clearly. They must try to speak in a similar manner once they start to employ VRS as a normal part of work. To that end, users need to avoid mumbling, hesitating, or slurring their words.
- VRS users keep the head and eyes in a fixed position for longer periods of time as they view the computer screen.
- VRS has been known to result in overuse of the voice. This can be combatted by taking regular breaks and by ensuring that users speak in a normal voice.
- Input devices should suit the shape and contours of the hands and fingers.
- The input device should move smoothly, be easy to control and operate, and should not need excessive effort.
- Any input device should allow for precise manipulation, respond promptly, and be capable of being used intuitively.
- Any buttons on an input device should be easy to reach and use.
- Left- and right-handed versions of input devices should be available for users so they can choose the appropriate model.
- The surface of the input device should be smooth and have no angled or sharp edges that can produce pressure points.

- Cables attached to an input device should not change the ease with which it can be used and moved.
- The response time between interaction of an input device and reaction on the screen should be such that the user does not perceive a delay.

References

Armand, J. T., Redick, T. S., and. Poulsen, J. R. 2014. Task-specific performance effects with different numeric keypad layouts. *Applied Ergonomics* 45(4), 917–922.

Armbruster, C., Sutter, C., and Ziefle, M. 2007. Notebook input devices put to the age test: The usability of trackpoint and touchpad for middle-aged adults. *Ergonomics* 50(3), 426–445.

Asakawa, D. S., Dennerlein, J. T., and Jindrich, D. L. 2017. Index finger and thumb kinematics and performance measurements for common touchscreen gestures. Gestures that can be completed more quickly may increase ease of use of touchscreens. *Applied Ergonomics* 58, 176–181.

Berolo, S., Wells, R. P., and Amick, B. C. 2011. Musculoskeletal symptoms among mobile hand-held device users and their relationship to device use: A preliminary study in a Canadian university population. *Applied Ergonomics* 42, 371–378.

Bridger, R. S. 2017. *Introduction to Human Factors and Ergonomics*. 4th ed. Boca Raton, FL: CRC Press.

Bufton, M. J., Marklin, R. W., Nagurka, M. L., and Simoneau, G. G. 2006. Effect of keyswitch design of desktop and notebook keyboards related to key stiffness and typing force. *Ergonomics* 49(10), 996–1012.

Cha, J.-M., Choi, E., and Lim, J. 2015. Virtual Sliding QWERTY: A new text entry method for smartwatches using Tap-N-Drag. *Applied Ergonomics* 51, 263–272.

Dennerlein, J. T., and Johnson, P. W. 2006. Changes in upper exptremity biomechanics across different mouse positions in a computer workstation. *Ergonomics* 49(14), 1456–1469.

Dennerlein, J. T. 2015. The state of ergonomics for mobile computing technology. *Work* 52, 269–277.

Forlines, C., Wigdor, D., Shen, C., and Balakrishnan, R. 2007. *Direct-Touch vs. Mouse Input for Tabletop Displays*. New York: Association for Computing Machinery.

Funch Lassen, C., Mikkelsen, S., Kryger, A. I., Brandt, L. P. A., Overgaard, E., Frølund Thomsen, J., Vilstrup, I., and Hviid Andersen, J. 2004. Elbow and wrist/hand symptoms among 6943 computer operations: A 1-year follow-up study (the NUDATA study). *American Journal of Industrial Medicine* 46, 521–533.

Gold, J. E., Driban, J. B., Thomas, N., Chakravarty, T., Channell, V., and Komaroff, E. 2012. Postures, typing strategies, and gender differences in mobile device usage: An observational study. *Applied Ergonomics* 43, 408–412.

Hsiao, H. C., Wu, F. G., and Chen, C. H. 2014. Design and evaluation of small, linear QWERTY keyboards. *Applied Ergonomics* 45(3), 655–662.

Jochems, N., Vetter, S., and Schlick, C. 2013. A comparative study of information input devices for aging computer users. *Behaviour & Information Technology* 32(9), 902–919.

Juul-Kristensen, B., Laursen, B., Pilegaard, M., and Jensen, B. R. 2004. Physical workload during use of speech recognition and traditional computer input devices. *Ergonomics* 47(2), 119–133.

Kang, H., and Shin, G. 2017. Effects of touch target location on performance and physical demands of computer touchscreen use. *Applied Ergonomics* 61, 159–167.

Kelaher, D., Nay, T., Lawrence, B., Lamar, S., and Sommerich, C. M. 2001. An investigation of the effects of touchpad location within a notebook computer. *Applied Ergonomics* 32, 101–110.

Kim, H., and Song, H. 2014. Evaluation of the safety and usability of touch gestures in operating in-vehicle information systems with visual occlusion. *Applied Ergonomics* 45, 789–798.

Kim, J. H., Aulck, L., Bartha, M. C., Harper, C. A., and Johnson, P. W. 2014. Differences in typing forces, muscle activity, comfort, and typing performance among virtual, notebook, and desktop keyboards. *Applied Ergonomics* 45, 1406–1413.

Kumar, R. 2008. A comparison of muscular activity involved in the use of two different types of computer mouse. *International Journal of Occupational Safety and Ergonomics (JOSE)* 14(3), 305–311.

Lin, M. Y. C., Young, J. G., and Dennerlein, J. T. 2015. Evaluating the effect of four different pointing device designs on upper extremity posture and muscle activity during mousing tasks. *Applied Ergonomics* 47, 259–264.

Mohs, C., Hurtienne, J., Scholz, D., and Rotting, M. 2006. Intuitivitat: Definierbar, beeinflussbar, überprüfbar!. In: Proc. Useware, pp. 215–224.

Noah, B., Li, J., and Rothrock, L. 2017. An evaluation of touchscreen versus keyboard/mouse interaction for large screen process control displays. *Applied Ergonomics* 64, 1–13.

Oney, S., Harrison, C., Ogan, A., and Wiese, J. 2013. ZoomBoard: A diminutive QWERTY soft keyboard using iterative zooming for ultra-small devices. In: Mandl. T (Ed.), *Proceedings Of the Special Interest Group on Computer–Human Interaction Conference on Human Factors in Computing Systems.* New York: ACM, pp. 2799–2802.

Orphanides, A. K., and Nam, C. S. 2017. Touchscreen interfaces in context: A systematic review of research into touchscreens across settings, populations, and implementations. *Applied Ergonomics* 61, 116–143.

Pereira, A., Lee, D. L., Sadeeshkumar, H., Laroche, C., Odell, D., and Rempel, D. 2013. The effect of keyboard key spacing on typing speed, error, usability, and biomechanics: Part 1. *Human Factors* 55, 557–566.

Pheasant, S., and Haslegrave, C. 2006. *Bodyspace: Anthropometry, Ergonomics and the Design of Work.* 3rd ed. Boca Raton, FL: CRC Press.

Rempela, D., Barra, A., Brafmana, D., and Young, E. 2007. The effect of six keyboard designs on wrist and forearm postures. *Applied Ergonomics* 38, 293–298.

Sandnes, F. E., and Aubert, A. 2007. Bimanual text entry using game controllers: Relying on users' spatial familiarity with QWERTY. *Interacting with Computers* 19, 140–150.

Schürmann, T., Binder, C., Janzarik, G., and Vogt, J. 2015. Movement transformation on multi-touch devices: Intuition or instructional preparation? *Applied Ergonomics* 50, 251–255.

Shin, G., and Zhu, X. 2011. User discomfort, work posture and muscle activity while using a touchscreen in a desktop PC setting. *Ergonomics* 54(8), 733–744.

Sutter, C., and Oehl, M. 2010. Of age effects and the role of psychomotor abilities and practice when using interaction devices. In: Marek, T., Karwowski, W., Rice, V. (Eds.), *Advances in Understanding Human Performance*. Boca Raton, FL: CRC Press/Taylor & Francis Group, pp. 757–766.

Sutter, C., and Ziefle, M. 2006. Psychomotor performance of input device users and optimized cursor control. In: *Human factors and ergonomics Society Annual Meeting Proceedings. Computer Systems*. vol. 5, pp. 742–746.

Trudeau, M. B., Young, J. G., Jindrich, D. L., Dennerlein, J. T. 2012. Thumb motor performance varies with thumb and wrist posture during single-handed mobile phone use. *Journal of Biomechanics* 45, 2349–2354.

Van Niekerk, S. M., Fourie, S. M., and Louw, Q. A. 2015. Postural dynamism during computer mouse and keyboard use: A pilot Study. *Applied Ergonomics* 50, 170–176.

Woods, V., Hastings, S., Buckle, P., and Haslam, R. 2002. Ergonomics of using a mouse or other non-keyboard input device. HSE Research Report 045. Sudbury, UK: HSE Books.

Woods, V., Hastings, S., Buckle, P., and Haslam, R. 2003. Development of non-keyboard input device checklists through assessments. *Applied Ergonomics* 34, 511–519.

Wu, C.-F., Lai, C.-C., and Liu, Y.-K. 2013. Investigation of the performance of trackpoint and touchpads with varied right and left buttons function locations. *Applied Ergonomics* 44, 312–320.

Wu, F.-G., and Luo, S. 2006. Performance study on touch-pens size in three screen tasks. *Applied Ergonomics* 37, 149–158.

Zecevic, A., Miller, D. I., and Harburn, K. 2000. An evaluation of the ergonomics of three computer keyboards. *Ergonomics* 43(1), 55–72.

7

Organisational Issues

7.1 Introduction

Organisational issues should demand a lot of attention for the very simple reason that if they are managed correctly, they will reduce the risk of an arduous task. However, if they are badly managed, this could increase the likelihood of individuals encountering difficulties. In office workers, effort, reward, over-commitment, and perceived stress were shown to be the workplace stressors most consistently related to neck and upper limb pain (Bongers et al., 2006; Eltayeb et al., 2009; Huysmans et al., 2012; McLean et al., 2010).

To allow a job to be performed without risk to the user, thought has to be given to a number of issues, such as the speed with which the task is performed, how long it is performed without interruption, how long a rest break should be, and whether there is variety in the work.

7.2 Job Design

Office work has undergone a radical change over recent years. Technological changes have led to a situation where users have little reason to leave their workstations. Today's office is managed in such a way that workers can access almost anything they need using their computer system. They rarely need to collect a reference document, because they can get what they need on the Internet or the company's intranet. They do not need to deal with mail or attend many meetings because they communicate by email—even with people in the same department. There are many types of businesses where workers do not have to deal with any documents—for example, call centres, where users manage incoming calls or interact with customers through live online chats.

From a business point of view, it makes sense to ensure that the workforce 'specialises' in particular fields. By specialising in one field, they become skilled at that particular task and faster. The only problem with this approach is that the task becomes monotonous and repetitive. As a consequence, many workers have started to experience discomfort and pain brought on by the repetitive nature of tasks, usually combined with a lack of rest breaks and a lack of task variety. When a job lacks variety, it ultimately means that the body, or small areas of the body, such as the hands or arms, are used repeatedly in the same way. This way of working is repeated throughout a working day and from day-to-day. As a consequence of this type of work, the musculoskeletal system can become overloaded, because it was not designed to be used robotically over extended, uninterrupted periods of time.

Many users can be considered to be performing a highly repetitive task given the frequency with which they depress keys on either a standard or virtual keyboard, click the mouse, or roll the scroll button on a mouse. Some incidental tasks that accompany the keying or mousing task may also be performed repetitively without this being noticed. For instance, if an individual refers to a document while keying the details into the system, they may rotate their head from side-to-side as they refer to the document and then look back at the screen. Users may repeatedly move their head between a number of screens. It is common for users to complain about neck pain, and, often the repetitive swinging of the head back and forth is overlooked as a possible contributor. Data entry operators usually key in at very rapid rates, with 25,000 depressions per hour not being uncommon. Although this element of their task is correctly identified as highly repetitive, their other repetitive task often goes unseen. Data entry operators usually refer to paper documents that contain the data they are inputting, and as they key the data in with one hand, often with the right on the number pad, they flick the paper documents over quickly with the left hand as they transfer each document from the 'to-do' pile to the 'done' pile. The repetitive gripping and turning of the documents is an integral and potentially hazardous element of their task and should be included in any consideration or assessment of the task.

Offering variety in the work is one way of reducing the risk of users experiencing symptoms of ill health. Ellegast et al. (2012) have shown that specific tasks performed strongly affects measured muscle activation, postures, and movements. In factory settings, operators who perform highly repetitive assembly-type operations would be offered job rotation, in which they move from one distinct task to another after a period of time, such as an hour or two. The aim of rotation is to offer them a break from a particular feature of one task that might be considered hazardous by moving them to a qualitatively different task where the previously identified hazardous element is absent. For this system to work effectively, a full understanding of all of the task demands is required. There are many tasks in factories that place very similar, if not identical, demands on the operators, and simply moving operators to different positions along the length of an assembly line will not guarantee that they will benefit from the change. Prior to any programme of rotation being developed, a full analysis of the tasks should be undertaken.

This may require a detailed analysis of the postures employed, both upper limb and whole body, the sequence of movements required, and the forces employed, as well as a breakdown of the cycle in terms of duration and repetition. To ensure that there are sufficient numbers of skilled operators to move between the tasks, they should be offered appropriate training in each of the tasks they may be expected to perform.

It is unlikely that a job rotation system would be employed in an office setting. The alternative approach that can be suggested in an office is to offer workers the opportunity of carrying out other tasks in addition to the work they already perform. This is referred to as 'job enlargement' and does not usually entail an increase in responsibility. If the work lends itself to this process, perhaps some workers could share the task of opening the mail rather than one individual carrying out this task. Perhaps filing could be shared out between users rather than having office juniors carry out all of this work. There are many tasks that have been relegated to the 'menial' category and that are carried out by a single individual within the office. Also, some businesses take this relegation approach because it is a more efficient system that ensures that workers remain at their desks, continuing to work without the distraction of doing other tasks. This whole area should be revisited and approached with a different perspective. To ensure that the process of enlargement works effectively, users should be briefed that they must use the additional tasks to interrupt their screen-based work at regular intervals. They should not view these additional tasks as something to get out of the way as quickly as possible so they can get on with their 'real' job, the screen-based work. They must understand the benefits to be gained from using these additional tasks to break up their screen-based work.

Job enrichment is a further option; it is different from job enlargement in that it gives workers greater responsibility for their own work. For instance, they might be able to make a greater contribution to how they feel they or their team should perform their own tasks. This approach creates a more positive attitude to work and reduces monotony.

7.3 Work Rate

It is not possible to state categorically what speed people can work at safely and comfortably in an office, apart from making a general comment that they should work well within their 'optimum' level of functioning. Given that people are so different, they will have their own individual levels of ability and, as a consequence, can potentially work at different speeds. However, most working environments impose expectations on employees in terms of what they have to achieve by the end of their working day.

To maintain the work rate within a zone that is considered acceptable, a number of matters needs to be addressed. The first of those is the issue of work hardening. It is generally considered the case that, over time, people develop a level of task fitness to meet the day-to-day requirements of their work. This takes a period of time, usually weeks, during which the person works consistently, eventually coping with the demands of the tasks without feeling particularly fatigued—assuming that the task demands were not excessive in the first instance. This process is identical to that which occurs when an individual takes up a new sport. Most people accept that it will take them some time to become accustomed to the demands of a new sport, and, after a period of exposure, they will feel that they can cope more easily with the demands of that sport. Because it takes time to become work-hardened, it is usually recommended that new recruits be introduced to new ways of working on a gradual basis. Allowing them an acclimatisation period makes them unlikely to be pushed to a point where their level of task fitness falls short of what is required to perform the task. If such a mismatch occurs, this is likely to be a source of difficulties for the workers. It is common for them to experience physical discomfort in such circumstances, particularly in the upper limbs.

A similar acclimatisation period is needed, ideally, when people are faced with an increase in workload or change in working circumstances generally. A sudden change in the way they have to work can be sufficient to trigger feelings of ill health. It is common for certain departments within organisations to be presented with cyclical peaks and troughs in workload. Accounting departments often find that at the month-end or year-end they have to prepare spreadsheets or printouts that require a lot of intensive keyboard and mouse work. These changing demands do not sit comfortably with the general view that it is desirable to have a level of demand that remains consistent from day to day and from week to week, avoiding the peaks and troughs that can cause sudden increases in work demands.

An individual's level of work hardening will decrease following an extended absence from work, such as through sick leave or maternity leave, and, once they return to work, they should not be expected to pick up exactly where they left off and get on with their work at the same pace as before their absence. Anyone returning to work after an absence should be permitted to return on a gradual basis so that they can build up to their previous level of output. Putting this in a sporting context again, if an individual had not played a sport, such as football, for weeks or even months, no one would find it hard to accept that they needed time to get re-accustomed to the sport before being able to play at the same level again. The same thought and consideration should be given to people in the workplace. This principle is also relevant when workers have extended holidays or the business closes down for a holiday period, such as at Christmas.

People who are responsible for determining their own way of working should be aware of the need to work consistently throughout the day. Few appreciate that working intensively at the start of their shift, followed by a more leisurely second part of the shift, actually has a more detrimental effect on them than working at a steady pace throughout the day. Again, putting it in a sporting context, most people would accept that if running in a marathon, they would be more likely to successfully complete the marathon and avoid adverse effects if they ran at a consistent pace from start to finish. Workers would readily accept that if a runner set off at a rapid rate at the start of a marathon, they would be unlikely to get very far before experiencing ill effects, such as extreme fatigue which, ultimately, might result in them failing to complete the race. This has parallels with the workplace. People will work more comfortably with more consistent quality if they maintain a steady pace throughout the whole of the working period. This clearly illustrates the need for awareness programmes for employees so that they understand what can cause them to experience discomfort or ill health and the need to develop suitable working practices.

7.4 Rest Breaks

Having established what work needs to be completed by an individual and within what timeframe, consideration needs to be given to the rest breaks. Decisions need to be made about how frequently an individual can take breaks and how long those breaks need to be to have a positive impact. In the past, many offices followed the traditional pattern of one midmorning break of about 15 minutes, a lunch break of between 30 minutes and one hour and a 15-minute break mid-afternoon. In effect, workers got a dedicated rest break about every two hours. Although some offices maintain this regime, many have abandoned it in the belief that workers are able to take breaks away from their desks any time they like, within reason, and therefore there is no reason to provide dedicated rest breaks. This approach needs to be monitored closely for a number of reasons. For instance, some office managers may have a perception that workers can leave their desks at least once a hour to collect a drink from a vending machine, thereby offering a rest break. However, it is common for groups of users to work as part of a team, and for each member of the team to take it in turn to collect the coffee for the remainder of the team members. If there are eight users in the team and they have coffee hourly, this could result in team members five, six and seven having to wait until lunchtime before they leave their desks, as team members one to four have collected the coffee at 9.00 am, 10.00 am, 11.00 am, and noon. Other businesses use monitoring systems in which workers have to record each time they leave their workstations by both logging off the system and/or manually completing a paper record. They often have to specify the reason for leaving the workstation. It is common for office managers to compare the computer-generated statistics based on the logging on/off times with expected standards to identify any discrepancies in times recorded and to look for explanations if there is a greater discrepancy than they deem acceptable. Such a system actively discourages individuals from leaving their workstations. This is also true of an overbearing management style in which an office manager or team leader glances at their watch each time an individual leaves and returns to the workstation in order to calculate the length of the break. If workers are not going to be provided with formal rest breaks, then the system in place should not actively discourage them from leaving their workstations at regular intervals. It has been shown (Sharan et al., 2011; Huysmans et al., 2012; Menendez et al., 2008; Norman et al., 2008) that workers with high levels of workplace stressors may reduce their computer break periods, increasing the risk of developing acute discomfort or long-term neck and upper limb pain.

It is generally accepted that if people perform a continuous screen-based task, they should be able to stop work at hourly intervals. This can be achieved by performing another, non-screen-based task or by having a dedicated rest break. Although many office jobs are heavily reliant on the use of computers, there are still a number of tasks that employees perform that could be used to interrupt their screen-based work—for instance, sorting and opening mail, making telephone calls, attending meetings, and filing. Workers need to understand the importance of managing their work so that they use incidental tasks to get a change in activity and a break from the computer. For those who do nothing other than interact with a computer system, there is no choice but to offer them additional dedicated rest breaks.

Although it may appear that some office personnel work in very negative environments where they are almost continually attached to their computers, it has to be recognised sometimes that users get more breaks than they realise, even if they are not permitted to take formal rest breaks. This aspect is particularly important in situations where risk assessments are being completed in respect of a user's workstation and task demands. An assessor may be told by a user that their work is 90–95% computer-based, which would be a concern, but with a little bit of exploration, the assessor may find that this is not the reality of the situation. For instance, it is common for users to share printers, which could result in them leaving their workstations several times an hour or many times a day. They may have to access paper files stored in another area of the office, which necessitates that they leave their workstations, both to collect the files and to return them. They may need to collect drinks from another area of the building. They may have to leave the building to smoke. They may attend meetings or team briefings every day or every week. In general, there are many additional activities to which people are exposed during the course of their work, which many of them do not recognise and whose benefits they therefore do not see. However, this does not overlook the fact that there are many individuals working in offices who have nothing to do other than interact with their computers, and these are the ones who require priority attention. It is worthy of note that the number of short computer breaks, lasting between 30 seconds and 5 minutes, was approximately 20% lower for those who were expected to make a lot of effort when working, compared to those who were expected to make limited effort (Eijckelhof et al., 2014).

When deciding on a system of breaks, decision makers should understand that what is important about break taking is the timing of the break. How long the break lasts is the secondary issue. Rest breaks should be taken before the onset of fatigue, not at the point when people are already starting to feel tired and when their performance is deteriorating. If the break is taken before the onset of fatigue, the individual can be considered to be resting. If people take a break after they have started to get tired, they are actually considered to be recuperating. Once they return to work after their break, the individual who took the pre-fatigue break will be able to pick up where they left off, thereby maintaining their performance at the same level, whereas the individual who took their break post-fatigue will find that they have to build back up to their peak level of performance. This shows that individuals who take more regular breaks, and at intervals that prevent the onset of fatigue, maintain their performance at a more consistent level throughout a day when compared with those who wait until they are tired. One problem is that, on a superficial level, some companies view break taking as non-value-added time and, as such, believe breaks do not contribute to business efficiency. However, this is a very short-sighted view. People who have short regular breaks are better performers than those who work longer before having a break. Hourly breaks, when the screen-based work is continuous, are considered appropriate. Breaks every two hours when there is variety in the work is also appropriate. Workers who find that they are under pressure on occasion to hit targets may decide to work through breaks to make up time and catch up. This is often combined with working a longer day, such as coming in earlier and finishing later. It is at times like this that they make themselves vulnerable to injury.

Some organisations enter into phases of negotiation with their workforce about the rest break allowances and often suggest options in terms of break taking. If people are expected to choose a work-rest schedule, it should be from a position of knowledge regarding what are considered suitable working practices. This has particular relevance when it comes to deciding on the length and timing of breaks. It is common for representatives of the workforce to negotiate for a total number of minutes in breaks per day and then to specify when these minutes can be taken. However, where this negotiation often fails the workforce is when they come up with the option of allowing people to work through one or more of their breaks so that they can leave earlier at the end of the day. All this achieves is users working with their computers for longer uninterrupted periods. Anyone making decisions about how long people will work and what system of breaks they will be permitted should have a thorough understanding of what is most likely to promote a safe system of work.

7.5 Overtime

It is not uncommon for office staff to work overtime. Obviously, overtime results in them remaining in the working environment for a longer time than their standard daily working hours. As a consequence, they are exposed to the same working conditions for an extended period, and if that environment contains adverse elements, it is likely that overtime will increase the risk of injury, or at least dissatisfaction. Although remaining at the office longer is the recognised impact of overtime, it is rare to recognise that overtime also results in a reduction in out-of-work hours and a reduction in potential rest time at home. Therefore, when considering the effects of overtime, employers need to include both aspects in the equation. Particular care needs to be taken with employees who have reported experiencing musculoskeletal symptoms, such as backache or upper limb discomfort. If they are exposed to working conditions that are likely to be provoking their symptoms, such as highly repetitive keyboard work or poor workstation design, then questions need to be asked about whether it is advisable for them to work overtime. Clearly, many businesses rely on their employees working overtime, but they should realise that this overtime needs to be monitored and managed properly. Care should be taken that the staff members take regular breaks during the period of overtime and, if they start to experience ill health as a result of working extended hours, thought should be given to stopping it.

7.6 Incentives

Incentives can take many forms. Employees can be paid an extra amount for every additional item of work they process. They can be offered treats, such as vouchers for particular stores or memberships in gyms, if they achieve their targets. Teams can be offered away-days or meals in restaurants based on their annual performance. They may simply be told during a team brief at the start of each shift who the best performer was the day before. All of these incentives are intended to increase the likelihood of users hitting their targets, which are set by the company and intended to increase business efficiency. The difficulty with offering any kind of incentive is that it can have a noticeable effect on the way an individual chooses to work. Frequently, incentives cause people to work faster or work longer hours. It is not uncommon for workers who have monthly or quarterly targets to change their pattern of work suddenly as the deadline approaches. They start earlier in the day, work through breaks, and remain in the office longer at the end of the day. Some may even take work home. It should not be surprising when such individuals start to complain about pain or discomfort or, generally, start to feel pressured and dissatisfied. The impact of incentives needs to be monitored carefully.

7.7 Motivation

Motivation is an internal state that develops within an individual and is experienced only by that person. It varies over time and across situations. It results in a desire to act and is mediated by an intention to act in a particular way. A person's state of motivation can be influenced by external factors, including other people. It was suggested by Maslow in the 1950s that motivation was the driving force behind behaviour. He wrote that people had five classes of needs, which could be arranged in a hierarchy. The most basic needs at the bottom of the hierarchy are physiological needs, including those for food, drink, and shelter. Once these needs are fulfilled, they are replaced by others higher up the hierarchy. The next level includes safety needs such as security and freedom from fear. The third level is described as social needs and includes acceptance, affection, and belonging. The fourth level is represented by self-esteem needs, such as achievement, recognition, and approval. The final level is self-actualisation, where Maslow believed people would reach their full potential. A fundamental belief behind Maslow's theory is that higher-level needs do not present themselves in the equation until the lower-level needs are satisfied. The effort to fulfil these needs drives behaviour. A single motive may result in a series of behaviours, whereas one act on a person's part may have many motives, each of which can vary in intensity.

Motivation can be manipulated by offering workers rewards, such as bonuses. This is referred to as extrinsic motivation, which is different from the intrinsic motivation experienced when people are interested in the work they are doing. It has been suggested that personality traits such as extroversion and introversion affect how successful incentives are at motivating people. Extroverts do better when rewarded for success, whereas introverts improve their performance more if they are penalised for failure. High achievers tend to be motivated simply by performing a demanding task. Low achievers are motivated more by external rewards. External rewards will change the way people view the tasks they are being asked to complete and influence the strategies they adopt to complete them. If an individual is being offered a reward to improve performance, they need to have an expectation that they are likely to succeed before they start. If they feel they are trying to achieve an impossible goal, they will be 'demotivated', regardless of the reward being offered. Offering a significant incentive perceived as having a high value (monetary or otherwise) can actually impair performance if the individual becomes anxious.

7.8 Shift Work

Shift work is not generally associated by many people with office work. People tend to view offices as operating traditional 9.00 am to 5.00 pm schedules. However, there are many options available to the workforce that allow them to work outside of the normal office hours, mainly supported by the use of flexi-time. There are also many office environments that operate shift systems so that they have 24-hour coverage, such as call centres, emergency services, breakdown services, and overnight secretarial services (known as 'night owls'). Shift work usually means that there is a handover of duty from one individual to another; this can occur in the afternoon, at night, or over a weekend. Shift workers in offices can work shifts of 12 hours or more, but many work the standard 8-hour shift. Shift workers also usually work a rotating shift pattern where they work on nights, followed by time off, followed by daytime work. A slight variation on the typical form of shift work involves those individuals who are required to be 'on call' throughout a night or weekend having already worked a full day.

It has long been recognised that workers who cover night-time hours or early mornings tend to experience a number of unpleasant side effects, including a disruption of the internal body clock. The internal body clock regulates the sequence of biological activities that have an impact on how the body functions at any particular time of the day. For instance, heart rate, body temperature, and blood pressure are higher during the day than at night. The cyclical increase and decrease of those elements is referred to as the circadian rhythm; it is influenced, to an extent, by external cues, both social and physical, such as regular mealtimes, regular bedtimes, external light levels, and patterns of behaviour such as regulated work times. Although the internal body clock will adapt to changes in working patterns, such as when an individual works through the night, it is unlikely that it will ever fully adjust to night-time working. If the worker has days off, such as over a weekend, and adopts the more normal pattern of getting up in the morning and going to bed in the evening, their body clock will try to reset itself. This undermines the previous adjustment phase, making it more difficult for the individual to cope with night-time work once they go back to work after the weekend. If an individual works a rotating schedule that alters the shift every week, they may find that they remain in an interim state of not adapting to suit any shift pattern, which can culminate in circadian dysrhythmia and very poor sleep patterns. Temporary workers who are not familiar with the prevalent shift system within an organisation may struggle to cope with shift work. Those who have worked a different system for another employer may find it even more difficult.

Sleep disturbance is a common feature of shift work. Some find it hard to go to sleep during the day. Others may find that their sleep is interrupted by other people making noise as they go about their daily lives, and because it is generally brighter. Daytime sleep does not appear to have the same restorative qualities as night-time sleep. This is mainly due to the fact that daytime sleep is often lighter and shorter than night time sleep and can be interrupted by noise. If an individual becomes sleep-deprived, there is likely to be an impact on performance, particularly in tasks involving decision making or monitoring. Reaction time and memory will also be adversely affected.

Sleep disturbance can lead to fatigue, as can disruption of the internal body clock. Fatigue can be described as a decline in mental functioning and physical performance. The situation can be compounded by a heavy or demanding workload. Chronic fatigue, which can result from badly managed rotation of shifts and excessive working hours, can culminate in serious long-term health problems, such as digestive disorders and heart disease, and a worsening of pre-existing health conditions, such as epilepsy, asthma, diabetes, and mental illness. Individuals with diabetes may be considered more vulnerable, generally, because taking their medication is time-dependent. Shift workers also are more vulnerable to minor ailments, such as colds and flu. To allow for an accurate diagnosis of shift-enhanced ill health, workers should be encouraged to inform their doctors that they work shifts. As a result of their poor sleeping pattern, some employees may start to rely on sleeping medication, and those who experience fatigue at work may start to rely on stimulants of varying types to remain alert. All of these factors are likely to have an impact on their performance and can be disastrous in safety-critical situations. Supervisors, generally, should understand and recognise signs of shift-related problems so that they can step in when necessary and take appropriate action to control or combat the consequences of such problems.

The overall impact of shift work is mediated by personal characteristics. For instance, some individuals find it easier than others to go to sleep and stay asleep during the day. Those with a history of sleep disorders are more likely to struggle with sleeping during the day. Workers who find that they need nine hours or more of sleep to feel refreshed are less likely to cope with shift work successfully than individuals who only need five hours of sleep or less. People who normally 'get up with the larks' find it more difficult to adapt to night-time work, as do people over 50 years old. Young workers and new and expectant mothers are also considered more vulnerable if included in shift work. Levels of general fitness appear to have an impact on the ability to adapt to changes in working pattern. Anyone with a history of drug or alcohol abuse or with a known mental health problem will be less successful at working shifts. Individuals who try to fit in a second job and those who have many household duties to deal with will also struggle to cope with changes in working patterns.

Those who are robust enough to cope with shift work will continue to work within such changing patterns of work, and those who are not will opt out of the upheaval caused by shift work by finding another job. Those who remain are likely to be individuals who are better at developing and adopting coping strategies that make it easier for them to adapt to working late nights. This might suggest that in order to assist the workforce to cope with working on shifts, they should be helped to develop suitable coping strategies. For example, rather than offering night workers stodgy meals such as chips and pies, employers should offer them lighter, tasty meals that are easier to digest. This will make the workers less likely to suffer indigestion as a result of trying to process heavy foods late at night when their systems would normally be suppressing food intake. Workers should be offered advice about the best times to eat and what sorts of food they should avoid. Employers should also ensure that the restaurant/canteen facilities offered to night workers are as good as those offered to employees working during the day. Employers might also need to consider increasing the temperature in the offices for night workers to combat the effects of the normal night time drop in body temperature. They should review transportation facilities and commuting times for their workforce and identify whether these factors play a part in increasing fatigue, particularly among early starters and late finishers who may find it hard to use public transport during 'out of service' hours.

Shift workers should be advised about other coping strategies, such as developing regular sleeping patterns. They should approach daytime sleep just as they would night time sleep and go through the same rituals, such as going to the bedroom to sleep rather than sleeping on the sofa, closing the curtains, cleaning their teeth, and changing into night clothes. In addition, they should switch off the ringers on telephones and introduce a system so that family members know that they are asleep and to avoid disturbing them. To ensure that they are ready for sleep once they get into bed, they should avoid drinking anything containing caffeine up to five hours before bedtime. They should not use alcohol as a means to hasten the onset of sleep, as this is likely to result in disruption of sleep. Workers should be advised to go to sleep as soon as possible after they return from their shift, rather than doing household tasks or watching television.

It has been established (Hayashi et al., 2004) that a 20-minute nap will maintain alertness and performance at higher levels while at the same time lowering levels of mental fatigue; in contrast, a 20-minute rest simply reduced subjective experiences of fatigue temporarily. Thus, employers might want to consider a more radical approach to combatting fatigue caused by shift work. However, this approach has to be taken with care. If workers nap for longer than 20 minutes, they can wake up feeling less alert and not refreshed. This may have an impact on their productivity and increase the risk of mistakes or accidents. It can take up to 15 minutes for these individuals to get back to their normal functioning level. This has implications in situations where the individual has to make decisions that are important or safety-critical. There are other practicalities associated with this approach. Appropriate rest facilities would need to be provided, along with an area where individuals could freshen up before starting back to work. Nappers would have to be supervised so that they did not inadvertently, or deliberately, sleep beyond the 20-minute allotted period.

It appears that weekly rotation of shifts, which involves individuals working between four and seven days before changing to another shift, is the most disruptive. This is because it does not offer sufficient time for the body clock to adjust but presents plenty of opportunity for sleep deprivation to amass. The extremes of rotation appear to be more successful. Working for one or two days before changing shift or working three weeks or more appear to be better at reducing the number of difficulties encountered by the workforce. It is also thought that a shift system that moves in the clockwise direction—so that the individual works the morning shift, followed by the evening shift, followed by the night shift—is more likely to result in a successful adaptation by the workforce than a counter clockwise system in which the individual works nights, then evenings, then days.

Shifts of eight hours are considered to be the most likely to result in consistently better work. For that reason, it is recommended that individuals performing safety-critical tasks, or tasks requiring concentration and alertness, should work only eight-hour shifts. Twelve-hour shifts offer benefits to the workforce in that they will probably work a shorter week. However, their fatigue levels are likely to be higher at the end of the shift and during the night, which will have an impact on performance. Limiting 12-hour shifts to two or three consecutive nights is likely to limit the build-up of fatigue. People working a standard eight-hour shift should not have to work more than seven consecutive days before having time off. As workers are likely to become more vulnerable to illness if they work in excess of 12 hours per shift, the amount of overtime performed should be limited. Working in excess of 12 hours is usually accompanied by a deterioration in alertness and accuracy, which is likely to result in reduced quality of work. If working shifts in excess of eight hours, particularly night shifts and shifts that start before 7:00 am, people should not be required to work more than two to three consecutive shifts. Split shifts, where an individual's daily work is divided into two separate shifts, result in increased fatigue, because this lengthens the overall working day. Fatigue is heightened in situations where people have insufficient time to return home to rest between the two parts of the shift. There should be sufficient time between shifts to allow an individual to travel home, eat properly, sleep for a suitable length of time, and be involved in some family and social activities. It is advisable for a shift worker to be allowed two full nights' sleep when changing from day to night shift or night to day shift. Two days of full rest are generally considered necessary for full recovery from a shift in excess of eight hours at night or with a very early (pre-7:00 am) start before moving to a day shift. Shift workers should not be required to work overtime. To ensure that they do not feel obliged to do so, systems should be established to cover absenteeism, changes in workload, and any emergency situations.

It has been suggested that, if it is not possible to control the performance effects of shift work by ensuring that workers get sufficient uninterrupted sleep or by employing them on a particular shift for long enough, the employer should consider how work is distributed relative to the time of day. Because memory and reaction time are adversely affected by lack of sleep, perhaps tasks reliant on these aspects should be performed during the day. This has implications in terms of training and instruction intended for shift workers. Dangerous or safety-critical work should be performed during the day if possible. Simpler tasks can then be performed at night. To maintain alertness, there should be variety in the work; alternating between sedentary tasks and more physically demanding tasks will be the most effective schedule. If there is no choice but to have workers perform work that might be considered dangerous or critical during times when they are likely to be sleepy, such as at night, they need to be made fully aware of the risks associated with their particular shift pattern and the implications of errors or misjudgements. Lone workers should be encouraged to develop a system of reporting to other workers at regular intervals through their shift, and, if they are working remotely from the main office site, they should have the communication facilities to do so.

As a general overriding principle, shift structures should be re-evaluated and modified to optimise workers' sleep opportunities (Vincent et al., 2016).

7.9 Compressed Working Week

A number of organisations that operate a 36-hour working week have considered whether they should permit their employees to work nine-hour days. This would result in the employees working a four-day week and having three days off. For the workforce, this sounds like an ideal solution for getting the balance right between home and work life. It appears that office workers do not display any adverse effects from working a nine-hour day, perhaps because they are not performing physically arduous tasks. As might be expected, office workers are more tired at the end of a nine-hour day than after an eight-hour day, but there does not appear to be any reduction in the quality of their performance. Having three days off appears to allow full recuperation from the effects of the longer days. Generally, levels of satisfaction are higher among individuals working a compressed working week owing to the fact that they have an extra day off work. From the company's perspective, it does not seem that productivity will increase as a result of this approach to the working week, and it has been stressed that identical responses to a compressed working week will not necessarily be replicated in other professions or other types of working environment (Josten, 2002). Generally, employers are advised against introducing nine-hour days where the workload is emotionally or physically demanding, where the employees have little choice over whether they wish to work a compressed working week, where an employee is not in good health, and where it is likely that the employees will have to work more than four consecutive nine-hour days.

7.10 Managing Change

Over the last 10 years, there has been a significant change in the physical spaces in which office workers find themselves. Prior to that most workers were bound to their office and their desk, simply because what they needed to do their work was located in one place. Mobile technology has enabled a change to take place in terms of the physical spaces where work is carried out. This, combined with the high cost of office space, has created a desire to encourage remote working and/or a more flexible use of office space.

One of the ways that offices are used more flexibly is through hot-desking. However, it has been recognised (Morrison and Macky, 2017) that this places additional demands and increased load on workers by creating an unfavourable physical working environment. This is referred to as 'Indoor Environment Quality' (IEQ) by Kim and de Dear (2013). They believe that this becomes detrimental to the individual located within it through reduced privacy, increased social distraction, and negative or emotionally demanding interactions with others.

Clearly, some ground rules are needed if hot-desking is to work successfully, although some common approaches do not always provide the intended outcomes. Brown (2009) showed that policies demanding clear desks and banning the personalisation of work spaces have little obvious benefit, and can, potentially, have quite negative consequences for some workers. These policies should be reviewed, as it now seems to be the case that the personalisation of workspaces fulfils some quite basic human needs.

Other changes that can occur within an organisation relate to its restructuring. A structure is needed so that it functions effectively. It determines the division of work and how all of the activities performed by the workforce are coordinated. When an organisation restructures, it has to examine the vertical and horizontal divisions within the company. Vertical divisions relate to the hierarchy within the organisation and are categorised by the degree of autonomy and authority associated with a role. Horizontal divisions relate to how one job differs from another and how each fits into a particular department.

A number of different forces result in organisational change: rapid changes in technology, competition, economic crashes, political climate, and social trends—for instance, more female workers in an environment that was previously male-dominated. These forces prompt the acquisition of vast amounts of new information and associated training, and previous knowledge and training may become redundant. There may be variations in patterns of work and, sometimes, even relocations in the business. Such changes suggest that a flexible workforce is desirable. To be introduced successfully, changes occurring as a result of restructuring have to be planned and all likely responses or outcomes should be anticipated. For instance, change brings with it uncertainty. Employees may fear the unknown and be unsettled by the knowledge that they need to acquire a new set of skills. This threatens the expertise they may have built up over years. They may need to form new relationships with people they have not worked with before. Some workers do not like to change well-established patterns of work and do not want to make the extra effort required to learn a new system of work—unless financially compensated. Some workers may resist change because they are complacent and do not believe there is a need for change; others may be concerned that they will lose power following a restructuring of the organisation or may lose their jobs altogether. On the whole, people resist change because of the personal loss they believe accompanies it.

For any change to be accepted successfully, one key point to be acknowledged is that, generally, people do not change themselves; they are influenced by others to change. As a consequence, they are more likely to accept change if they understand what is about to occur, if they believe they have had some input, if previous changes have been introduced successfully, if the consequences of the changes are understood and predictable, if the process of change is introduced through combined planning, and if there is overt support from higher management.

Reactions to change may vary significantly. The immediate reaction of some may be to resign from the organisation. Others may show active resistance to the change by refusing to accept it or by attempting to modify its form. In addition to showing personal defiance, these individuals may encourage others to follow their lead and resist also by, for example, going out on strike. Alternately, there may be group inertia where the whole group is obstructive and refuses to cooperate as a means to reinforce the position of the individual who shows resistance. Once a group is involved, it is difficult for individuals to adhere to the changes, because they could attract the disapproval of the group and possibly some form of penalty from it. A less extreme form of resistance may be demonstrated by some who employ behaviours that show their opposition to the change, perhaps through tactics that slow down the process of change, such as withholding information that is required prior to the change occurring. Such resisters do not appear to be overtly opposing the changes, but their behaviour speaks volumes. Other people affected by the change process may accept it passively, even though they may not be happy about it, because they believe it to be inevitable. This cannot be viewed as creating a positive working atmosphere. Others may have a more positive acceptance of the changes, while at the same time they have some reservations. These individuals may spend some time juggling the pros and cons of the changes in their own minds. More positive attitudes are shown by individuals who either accept the changes or actively support them.

There are a number of ways of reducing resistance to change. As change is more likely to be successfully introduced when people understand the process, educating and communicating with the workforce is desirable. Although this can be time-consuming, requiring the input of a number of individuals within the organisation, it often results in the workforce throwing their weight behind the change having been persuaded of its merits. Asking these same people to participate in generating the change package is also likely to be beneficial because it encourages commitment. This also provides an opportunity for the change agents to gather more pertinent information relating to the changes from the individuals who will be affected most during the process. This process can unfortunately prove time-consuming if the workers provide unhelpful or inaccurate information. Should people still resist change because they are not adjusting well to it, the organisation can offer support. If the resistance is born out of a recognition that an individual or group will lose out as a result of the changes, they can be encouraged to accept them by being offered some form of negotiated settlement. However, this sets a precedent that may be expected during any future changes, and this may be very expensive for the company. Finally, if companies are in a situation where time is running out and they have tried all other options, they can attempt to combat resistance to change through manipulation of a situation or coercion. Both of these can backfire on the company, leaving them discredited in the eyes of their workforce.

Part of the success of introducing change is using the right people to push through the changes. In-house staff and people from outside the organisation can all contribute to the change process; however, each can make either a positive or a negative contribution. Using in-house personnel to manage the change process means that the change agents have a more accurate view of how the organisation functions; they are closer at hand so are more readily available; they are known to the workforce; they have more control and authority within the organisation; and there are generally lower costs associated with their use. However, as an integral part of the organisation, they may not see what is going on right under their noses; they may have biased opinions of what is right for the organisation and the workforce; and they may create resistance if they are not perceived to be impartial. Independent change agents tend to have a more objective view of the organisation and have more experience in dealing with a wide range of potential problems, having encountered them in other organisations previously; and they are likely to have contacts they can call on for support or additional input. The downside of using external change agents is that they do not have a detailed working knowledge of the organisation, are unknowns, and have higher associated costs.

7.11 Summary

- Office workers should have variety in their work.
- Repetitive work can result in the development of upper limb disorders if sufficient rest breaks are not available.
- Job enlargement results in more tasks being performed without an increase in responsibility.
- Job enrichment gives workers greater responsibility for their own work, which increases satisfaction and reduces monotony.
- Workers need an acclimatisation period to become accustomed to their work demands and to be considered work-hardened or task-fit.
- An acclimatisation period is needed if the workload or working conditions are changed and following an extended absence from work.
- Employees should work consistently throughout the day and from day to day, and avoid peaks and troughs in activity.
- Computer users need regular breaks from their screen-based work. It is usually recommended that they have a break of about five minutes every hour.
- The system of monitoring employees should not discourage them from taking regular breaks.
- Breaks from screen-based work can include performing other non-screen-based work or taking a rest break.
- Breaks should occur before the onset of fatigue.
- Overtime not only exposes the individuals to their working environment for longer, it also reduces out-of-work resting time.
- Employers should ensure that regular breaks are taken during overtime hours.
- Employees experiencing work-related ill health should not work overtime.
- Incentives can adversely affect the way in which people work and should be monitored.
- Motivation is an internal state that can be influenced by external factors, such as other people.
- Motivation can be manipulated by offering rewards. The rewards will influence the strategies adopted to complete the work. Some strategies may have adverse effects.

- Working night shifts can disrupt the internal body clock and have an impact on biological activities and body functions.
- It is unlikely that the internal body clock will fully adjust to night time working.
- A rotating schedule of shifts that alters every week may result in the individual not adapting to suit any shift pattern.
- Sleep disturbance is a common feature of shift work. This can have an adverse effect on reaction time and memory.
- Chronic fatigue resulting from badly managed rotation of shifts can result in serious long-term health problems.
- Problems encountered by individuals working shifts over long periods include digestive disorders, heart problems, and worsening of pre-existing conditions such as epilepsy, asthma, diabetes and mental health problems.
- Personal characteristics play a part in how well a person copes with shift work.
- Employers should assist shift workers to develop coping strategies.
- Weekly rotation of shifts is considered to be the most disruptive schedule.
- Working one or two days, or three weeks or more, before changing shift is more successful in terms of adapting to shift work.
- A shift system that moves in a clockwise direction where the individual works a morning shift followed by an evening shift and then followed by a night shift will result in better adaptation than a counter-clockwise system.
- Shifts of eight hours are likely to result in consistently better work.
- Shifts of 12 hours should be limited to two or three consecutive nights to avoid the build-up of fatigue.
- Consideration should be given to assigning certain types of work to certain times of the day depending on the shift worker's level of fatigue and alertness.
- Compressed working weeks result in higher satisfaction among office workers.
- Employees fear the unknown and may resist change. They are more likely to accept change if they understand what is about to occur, they are involved, there is support, and previous changes have been successful.
- Communication with the workforce prior to and during the change will reduce resistance to change.

References

Bongers, P. M., Ijmker, S., Van den Heuvel, S., and Blatter, B. M. 2006. Epidemiology of work related neck and upper limb problems: Psychosocial and personal risk factors (part I) and effective interventions from a bio behavioural perspective (part II). *Journal of Occupational Rehabilitation* 16, 279–302.

Brown, G. 2009. Claiming a corner at work: Measuring employee territoriality in their workspaces. *Journal of Environmental Psychology* 29(1), 44–52.

Eijckelhof, B. H. W., Huysmans, M. A., Blatter, B. M., Leider, P. C., Johnson, P. W., van Dieen, J. H., Dennerlein, J. T. and van der Beek, A. J. 2014. Office workers' computer use patterns are associated with workplace stressors. *Applied Ergonomics* 45, 1660–1667.

Ellegast, R. P., Kraft, K., Groenesteijn, L., Krause, F., Berger, H., and Vink, P. 2012. Comparison of four specific dynamic office chairs with a conventional office chair: Impact upon muscle activation, physical activity and posture. *Applied Ergonomics* 43, 296–307.

Eltayeb, S., Staal, J. B., Hassan, A., and de Bie, R. A. Work related risk factors for neck, shoulder and arms complaints: A cohort study among Dutch computer office workers. *Journal of Occupational Rehabilitation* 19, 315–322.

Hayashi, M., Chikazawa, Y., and Hori, T. 2004. Short nap versus short rest: Recuperative effects during VDT work. *Ergonomics* 47(14), 1549–1560.

Huysmans, M. A., Ijmker, S., Blatter, B. M., Knol, D. L., van Mechelen, W., Bongers, P. M., and van der Beek, A. J. 2012. The relative contribution of work exposure, leisure time exposure, and individual characteristics in the onset of arm-wrist-hand and neck-shoulder symptoms among office workers. *International Archives of Occupational and Environmental Health* 85, 651–666.

Josten, E. 2002. *The Effects of Extended Workdays*. Amsterdam, the Netherlands: Royal Van Gorcum.

Kim, J., and de Dear, R. 2013. Workspace satisfaction: The privacy-communication tradeoff in open plan offices. *Journal of Environmental Psychology* 36, 18–26.

McLean, S. M., May, S., Klaber-Moffett, J., Sharp, D. M., and Gardiner, E. 2010. Risk factors for the onset of non-specific neck pain: A systematic review. *Journal of Epidemiology and Community Health* 64, 565–572.

Menendez, C. C., Amick, B. C., Chang, C. H. J., Dennerlein, J. T., Harrist, R. B., Jenkins, M., Robertson, M., and Katz, J. N. 2008. Computer use patterns associated with upper extremity musculoskeletal symptoms. *Journal of Occupational Rehabilitation* 18, 166–174.

Morrison, R. L., and Macky, K. A. 2017. The demands and resources arising from shared office spaces. *Applied Ergonomics* 60, 103–115.

Norman, K., Floderus, B., Hagman, M., Toomingas, A., and Tornqvist, E. W. 2008. Musculoskeletal symptoms in relation to work exposures at call centre companies in Sweden. *Work* 30, 201–214.

Sharan, D., Parijat, P., Sasidharan, A. P., Ranganathan, R., Mohandoss, M., and Jose, J. 2011. Workstyle risk factors for work related musculoskeletal symptoms among computer professionals in India. *Journal of Occupational Rehabilitation* 21, 520–525.

Vincent, G. E., Aisbetta, B., Hall, S. J., and Ferguson, S. A. 2016. Fighting fire and fatigue: Sleep quantity and quality during multi-day wildfire suppression. *Ergonomics* 59(7), 932–940.

8

Training

8.1 Introduction

This chapter is not intended to cover the topic of training in terms of an individual's professional development within an organisation. It deals with the need to train people to work safely, comfortably, and happily within their working environment.

For some reason, when it comes to working in offices, it is often assumed that people will know how to use desks and chairs and will know the best ways of working. There is a common belief that because people have been sitting on seats since they were babies, they must know how to sit. Yet, why would anyone know how to use three or four adjustment mechanisms on a chair to adopt a suitable posture relative to the desk, keyboard, and mouse, unless they were shown? They will not get divine inspiration; the knowledge will not suddenly present itself inside their heads. People need to be given specific and detailed information and training relating to the use of their workstation equipment and how they should work safely—not just how they do the job. This information needs to be presented in a format that is easy to understand, that relates to their own workstations and working environment, and that they will remember.

Training and information are needed not only to advise people about what they should be doing when at work in order to avoid premature fatigue, injury, or dissatisfaction; training should also draw their attention to the likely consequences of working in unsuitable postures and the possible adverse effects of bad habits. People with better health knowledge are more likely to choose healthy lifestyles. The positive correlation between health knowledge, attitude and behaviour has been well examined and documented in many studies (e.g. Trichesa and Giuglianib 2005).

The aim of the training should be to ensure that the workforce avoid injury or ill health as a result of their work or the way they do it. The training should be part of an explicit intention to develop a more positive attitude to health, safety, and well-being in the workplace, and the aim should be for this attitude to become second nature. If the training is successful, the employees will feel more competent in contributing to risk reduction strategies in the workplace. As a natural by-product, effective training should also reduce the costs associated with ill health and absences resulting from it. The success of training is, in part, influenced by the company's perception of it. For instance, does the company take the approach that training is an investment, or does it believe that it represents simply a cost to the company? If the latter is the case, training will often be under-resourced and competency may not be viewed as an ongoing issue needing regular updating.

Training is the means by which an individual acquires a set of rules, concepts, skills and attitudes. It is generally intended that training will lead to an improvement in job performance. However, training should also be viewed as a means to avoid work-related ill health. Whatever form the training takes, it should aim to achieve three things: the acquisition of knowledge, the creation of procedures and practices, and the development of an appropriate skill base. Table 8.1 shows a simple but accurate feedback loop that can guide through the development and implementation of any training. Although a way of ensuring that training works effectively was suggested by Lambert back in 1993, the idea that it should be 'double SMART' is still valid today (See Table 8.2).

TABLE 8.1

Feedback Loop for Improving Training Procedures

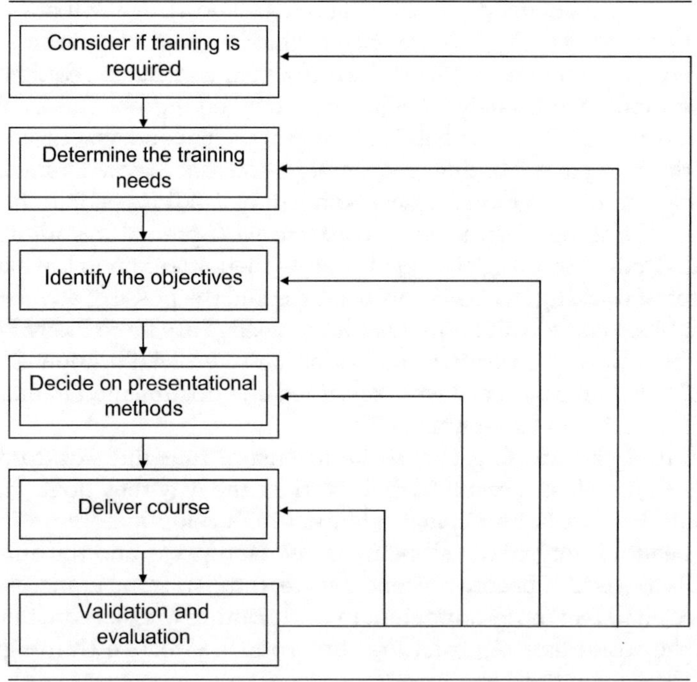

TABLE 8.2

The Conditions Under Which Training Works Best

S	*Specific* in that it should define and demand observable behaviours. *Significant* in that it makes a valid contribution to the desired culture, vision, mission, and goals of the organisation.
M	*Measurable* in that it can be applied over time consistently and constantly in all appropriate situations. *Meaningful* in that it is congruent with the values and proper expectations of the individuals who must make it work.
A	*Achievable* by the participants with further support. *Attainable* within the organisational, economic, legislative, and social climates as it exists.
R	*Realistic* in that adequate time is designed into the programme for sufficient threat-free practice by all. *Reward-driven* within the existing or emerging culture, so that desired behaviours will be reinforced in activity.
T	*Timely* so that it meets the credible present and future needs of the organisation. *Team-orientated.*

Irrespective of how well thought out a training programme might be, it should not be used, or viewed, as a substitute for proper risk control. For instance, if employees are given non-adjustable chairs to use when sitting at their desks, no amount of training will overcome the impact of the chairs' poor design.

8.2 The Trainees

At the outset of the programme, consideration should be given to who needs to be trained. Although the initial concern may be that the general workforce needs to be educated in appropriate posture and workstation arrangements, managers and supervisors also need to know what is required if they are to monitor the situation after the training has occurred. In addition, they should lead by example, and they can only do this if trained themselves. Training offered to supervisors and managers should include additional information relating to their responsibilities with regard to health and well-being in the working environment, and they should understand how the training and its intended outcomes fit into the overall company approach to health and safety. Many of them will be ignorant of not only the company's expectations of them, but also of the legal requirements. Additional advice on hazard identification and risk reduction strategies would be of benefit to them.

Some trainees have a preferred learning style. If this 'fits' with the preferred teaching style of a trainer or the organisation as a whole, the training is likely to be more successful. One view is that there are four learning types: activists, reflectors, theorists, and pragmatists.

Activists are happy to encounter new experiences. They are not sceptical, but rather are open-minded and quite enthusiastic about all things new. They thrive on on-the-spot problem-solving and do not resist change. Unfortunately, they tend to rush into change without due consideration and have a tendency to become bored after the initial enthusiasm has waned.

Reflectors prefer to stand back and evaluate what is occurring around them. They draw together information from many sources before deciding exactly what course of action they should take. As a consequence of steadily collecting information and analyzing it, these individuals are slow to make decisions about a course of action—that is, they delay the inevitable. These are the individuals who sit at the back of the training room and observe but not necessarily contribute until they are certain of everyone else's position on the training content. They are also less likely to offer the trainer an insight into where their concerns lie, because they are unlikely to be very forthcoming.

Theorists tackle all learning situations in a very logical way. They can pull together different arguments to reach a single conclusion. They analyse everything and reject any suggestions that do not offer certainty. They are likely to ask detailed questions during training and take a disciplined approach to the course. They are, however, intolerant of ambiguous messages during the training.

Pragmatists are willing to try out new approaches and techniques to test whether they work in the real world. They leave a training course full of enthusiasm and try out all the suggestions made during the course. They tend to be practical, problem-solving individuals who enjoy the challenge of trying something new. They are likely to reject any training that they feel is purely academic with no real application.

To summarise people's preferred learning styles, they can be divided into active and passive types. Active types learn by using their initiative, actually getting on and doing what they are supposed to do and exploring and testing new ideas. Passive individuals learn by observing, asking questions, imposing an interpretation on what they hear and see and reviewing their position on situations. Different types of learning rely on different learning methods. Table 8.3 provides insight into the types of learning methods employed by each style of training.

It is believed that people of low self-esteem are more easily persuaded by the arguments put forward by a trainer than are those of high self-esteem. Connected to this issue is the desire for social approval. Those who have a deep-rooted need for social approval are more likely to be

TABLE 8.3

Types of Learning Methods Employed by Each Style of Training

Training Type	Learning Method
One-to-one coaching	Doing
Online systems	Using initiative
Practical sessions	Using initiative/testing
Initial induction	Listening/questioning
Job shadowing	Observation/testing
Mentoring	Exploring
Networking/seminars	Listening/questioning
Long distance learning	Using initiative/exploring
Evening classes	Listening/questioning
Practical projects	Exploring/doing
Competitions	Using initiative/exploring
Workshops	Reviewing/doing
Mock-ups	Doing

influenced by what others do, such as whether they agree with the drive to change working behaviours. It also seems that if people are forewarned about the likely approach of the course, and how it might question their current beliefs about working, this will make them easier to persuade during the course.

Consideration needs to be given to any special requirements of those likely to attend the course—for instance, whether any have problems with vision or hearing, or have learning difficulties. All need to be catered to satisfactorily, and this can be achieved, but only if the trainer is aware in advance of the course that they will be attending.

All course participants should sign an attendance sheet, which should be stored for future reference. Although it may seem obvious, when filling in the attendance sheet each trainee should print their name clearly before signing alongside it. This is particularly important for the company, because it avoids any disputes in the future as to whether a particular individual did attend a course.

If there are many people within the organisation requiring training, they will need to be prioritised in some way. It is probably a good starting point to ensure that all new recruits are given the training as part of their induction at the outset of their employment. Those who have complained about encountering problems or those who have been identified as being particularly at risk, possibly during a risk assessment process or through general observations, should be included in the first batch of training. People who are taking on different and more challenging aspects of work should also be given priority.

Using personal development plans is useful within an organisation. This allows employees to define their own personal training needs and to specify how these might be achieved. This can assist in tailoring training to suit the individual's specific needs.

8.3 Training Methods

Training can address all the topics of interest in one session, or a series of sessions can address parts of what an individual has to assimilate. The latter approach ensures that people are not overloaded by the need to learn simultaneously about using their chairs, the layout of their desks, the use of their equipment, the need to take breaks, the most suitable way of managing their work, and so on. It has been suggested (Kroemer and Grandjean 1997) that having rest breaks during the session will have a dramatic effect on learning a skill. Frequent short breaks will result in more rapid acquisition of skills than if the session is continuous and without breaks. It has been suggested that the breaks allow for 'mental training', whereby trainees can think about what they are being asked to do.

The information needed to allow people to work comfortably and safely can be presented in a number of formats. It can be given to them during a formal training session that takes the form of a lecture in an area remote from their work area. They can have one-to-one coaching at their own workstations, they can watch videos in a training room, or they can engage in online training at their own desks. Each of these approaches has its drawbacks, which an organisation must recognise, so that it can avert any adverse outcomes.

In a lecture situation, there is typically a one-way flow of information as the trainer talks to the attendees. Attendees can sit passively throughout such a session and may not understand what is being covered unless the trainer specifically probes their understanding, either through questions, a test paper, or a practical demonstration. A trainer has to accept that many of the participants have not chosen to be there, but instead have been required to attend by their manager. As a consequence, they may have no interest in the proceedings and may spend the majority of the training period thinking about how much working time they are losing. As a consequence, they may not absorb what is being discussed. Group training such as this may not address individual needs, and some people might find it difficult to transfer what they have heard in a remote setting, such as a training room, to their own workstations where they have to apply the practical aspect of their training. However, training of large groups of people is time- and cost-efficient and does allow for information to be passed to many individuals in a short time.

The success of the course will be largely determined by the skill of the trainer, which has implications in situations where organisations simply decide to use one of their own in-house personnel, probably skilled in other areas, to pass on the relevant information. If in-house trainers are to succeed, they should have training in presentation skills. They should be able to generate a rapport with their trainees, be able to deal with difficult or generally argumentative individuals, and be able to deal with all styles of learning and types of learning difficulties. They should be capable of designing an interesting, effective session and be aware of the need to evaluate the trainees as they perform their practical skills and offer suitable feedback. They should be keen to evaluate the effectiveness of their course upon completion and be able to deal with further queries and concerns among trainees following completion of the course, and continue with an analysis of what further training is required. The company does need to be alert to the possibility that its own in-house trainer, who might be training as a secondary responsibility, may actually pass on poor working practices of which they themselves may not be aware. Sending in-house trainers on a 'train-the-trainer' course might be beneficial. On the positive side, using in-house personnel to present training courses does allow for the training to be scheduled to suit the business rather than fitting in with the schedule commitments of an external trainer. In addition, actually holding training in-house reduces costs because employees are not off-site for longer than is necessary, and there are no additional travelling costs for transporting the employees to the venue. It should be kept in mind, however, that employees are not always comfortable with being completely honest about their concerns and difficulties if they are being trained by someone they work alongside every day.

In their defence, external trainers are specialists in their fields; they are usually up to date on current legislation and industry standards and practice; and employees may feel more comfortable about being open with them.

Two important factors determine how persuasive a trainer can be when trying to communicate with a group about the need to change the way they work: the trainer's credibility and 'attractiveness'. The trainer's credibility rating among a group of trainees is influenced by their level of expertise and how trustworthy they appear. The more credible and trustworthy they appear, the greater their impact on the group. If the trainer can talk through the material smoothly and quickly, rather than hesitantly and slowly, they will be perceived to be more credible. The trainer's perceived level of attractiveness is influenced by personal factors, such as appearance and overt personality. If they are considered attractive, the trainees will attribute positive personal characteristics to them. This is called the 'halo effect'.

8.4 The Course

The content of the course is driven by what the trainees need to achieve once it has been completed. A review of previously recorded injuries or ill health can indicate where particular problem areas are within an organisation. Risk assessments are also a good source for identifying what needs to be addressed and in what depth, and these assessments are also likely to document what training has already been used as a control measure for risk. Risk assessments are dealt with in Chapter 13. Employees themselves can offer insight into what they feel they lack in terms of knowledge about their working arrangements. A knowledge of legal requirements with regard to what training should be offered is essential.

If an independent trainer is not being used and the company decides to use one of their in-house personnel to provide training, this individual may be at a loss as to what materials they can use. This is particularly the case if they have not done this type of training before. There are many sources for useful training information. For instance, trade unions, trade associations, and employer associations often have advisory information for trainers to use and leaflets to hand out on courses. Local training or community colleges may have subject-specific libraries that can be accessed for ideas. Some training companies provide materials for use on courses presented by other individuals.

The nature of the material and the way in which it is presented during a training course has an impact on its effectiveness. What a trainer is trying to achieve during the course is a level of emotional arousal, because emotions influence rational mental processes. Trainers have to decide whether they simply provide a mild argument supporting the need for a change in working posture—for instance, stating that it is better for the person's back if they sit properly in the chair—or whether they use more forceful imagery to back up their arguments, such as describing the range of potential back complaints and the level of pain associated with them. The more concerned people are about the information being given to them, the more likely they are to act on it—as long as they are given explicit practical advice on how to take action. Vivid imagery is more effective than lots of statistics at persuading people to change the way they work. For instance, showing a series of photographs of the workforce displaying a range of extreme postures used every day as a normal part of the work will probably have more impact than presenting statistics detailing the number of complaints of backache in a department or the number of days off due to ill health.

Consideration has to be given to whether trainers should present simply a one-sided discussion during the course, or whether they should examine both the supporting and opposing sides of an argument. If the trainees hold the same views as the trainer and accept the need to change the way they work, a one-sided approach is effective. If the trainees hold an opposing view to the trainer's—for instance, thinking the course is a waste of time because they do not believe they need to change the way they work—a two-sided approach is likely to be more effective. In addition, people with little knowledge of a subject are swayed more easily by a one-sided argument, and those with a greater knowledge base are persuaded more readily by a two-sided argument. Finally, trainers need to prepare themselves for a more challenging course if there is a large discrepancy between the views of the trainees or their desire to improve the ways in which they work and the message that is being put across during training. The larger the discrepancy, the greater the challenge. However, the greater the credibility of the trainer, the more likely it is that this chasm will be spanned.

It is the responsibility of the trainer to ensure that trainees understand what is being said by questioning their comprehension throughout the session. The trainer should be capable of dealing equally well with all individuals within an organisation, whatever their position in the hierarchy, and of matching the course content and approach to the trainees' specific abilities and needs.

Ensuring that room-based training includes practical sessions, where the individuals can work with a desk and chair and arrange workstation equipment within the training suite, increases the effectiveness of the training, because it shows the connection between the training environment and the real work environment. It also allows the trainer to observe any difficulties specific individuals might be having with following the advice being given.

It is generally believed that we remember 10% of what we read, 20% of what we hear, 30% of what we see, 50% of what we see and hear, and 70% of what we do for ourselves. If trainees are expected to learn from reading, they will learn more if they spend time trying to recall the information rather than simply re-reading the text. They are more likely to remember material they read if it corresponds with previously held views, as opposed to undermining their views. They will also learn more from the written text if they read it in the relevant environment—for instance, when sitting at their desks—and then try to implement the advisory information straightaway.

Positive reinforcement or feedback during the practical session gives individuals an immediate understanding of their performance at that moment and allows them either to continue with their current practices or to change them with a view to improvement. Precise feedback allows individuals to 'tailor' the changes in their behaviour. The feedback needs to be compatible with individual levels of understanding and the point each has reached in the training. It should focus on those aspects of behaviour that are critical in ensuring that trainees can subsequently work comfortably. The overall value of the feedback, as perceived by the trainee, is influenced by who dispenses it. If the trainer is highly respected, their feedback will be perceived to be of great value. If the trainer is too strict with the trainees and takes a more negative approach when offering feedback, although they will achieve a degree of conformity, they will generate anxiety in the trainees and possibly acquiescence. Criticism can result in a change in a trainee's self-confidence and sense of worth. Careful consideration has to be given to when it is likely to be unnecessary to offer any further feedback, as premature cessation of feedback can result in degradation of positive behaviours.

Showing video clips as part of a training course can be helpful because it allows the course participants to see the types of postures they should adopt when working. They also get an appreciation of how they should adjust the settings of their chairs and how to arrange the layouts of their work surfaces. However, watching only a video without any other form of input from a trainer is likely to have a limited effect in changing the behaviour of individuals in the working environment. Some people find it difficult to draw a parallel between what they are viewing and their own workplace and so will be unable to transfer the information. Organisations should also be alert to the fact that off-the-shelf video clips are not usually made by companies that specialise in the field of ergonomics. They may be specialists in creating films, but that does not guarantee that they will produce a film that accurately relays all the steps that should be taken by the trainees in terms of arranging their workstation. Video clips need to be augmented by additional input from the trainer, who either offers additional information or draws attention to weaknesses in the film and provides an alternative view.

8.5 Alternative Approaches

The effectiveness of group training can be increased by follow-up, one-to-one coaching at the workstation. This allows the trainer to measure each individual's understanding of the basic concepts and to determine whether the trainee has been able to return to the workstation and employ the advice offered during the training session. It is not uncommon for individuals to return to their workstations after a training session and still be unable to adjust their own chairs correctly. Some people also find it difficult to speak out in a large group. One-to-one coaching is a non-threatening opportunity for people to ask questions about their own workstation arrangements and how these can be altered. The trainer can also discuss the trainee's specific work commitments so that they can develop the work routine that is least likely to result in fatigue or injury.

Many organisations are now selecting online training, also known as e-training, as an efficient means to 'cascade' information to their workforce. The drawbacks of traditional face-to-face training, which include the distance to training centres for remote workers, the inconvenient timing of training for some, training content, the pressures of the growing workforce, logistics, diminishing budgets for traditional training, and the constraints of schedules have made e-training an attractive alternative (Loh et al. 2013).

Zainab et al. (2017) have shown that the cost usually associated with traditional training can be reduced, which increases the opportunity to invest in other aspects of the business, which can help in improving overall performance.

Online training allows trainees to use the system in their own time to fit in with their work demands. It does not require the employment of a trainer. The training material remains standardised over time. It does not require that large groups be made available at any one time, which can create problems in terms of keeping a business ticking over. The difficulties are that it is up to the manager of a group of employees to keep reminding them that they must work through the online training, which can sometimes be moved to the back burner given their other work demands. And, unless the system has some in-built means to test their level of knowledge, individuals can passively work through the package without absorbing the salient information. It is also thought (Wang et al. 2014) that most of the software packages intended to deliver online training either contain only limited information or are instruction-oriented, rather than motivational or encouraging.

The most successful online training packages are designed to test a trainee's knowledge base at intervals and do not permit them to proceed to the next phase of the training if they do not demonstrate an acceptable level of knowledge. Again, there is no guarantee with this system that individuals will be able to transfer what they read on the screen to their own working environment. Only a one-to-one follow-up can do this properly. Online systems also do not provide opportunities for trainees to ask specific questions relevant to their own workspace.

Informational leaflets are commonly used to advise computer users about the risks associated with their work. These, and the online systems available, such as the company intranet, form what could be referred to as 'knowledge banks'. Leaflets often detail the postures and working practices that should be adopted when working. However, these leaflets are often distributed away from the work area, such as during a meeting, and may not make it back to the work area or may have no immediate relevance for the individual. Many people will not read them at all, or, if they do, will not remember the content for long. Some authors (e.g. Visschers et al. 2004) have investigated the possibility of using warnings as an alternative. They concluded that it was possible to bring about 'on-the-spot position adjustments' in an office setting, such as the immediate changes that would occur in people's behaviour in a situation where they were warned to stay away from high-voltage electricity. They concluded that warnings could be provided more directly, would have brief but clear content, and would attract more attention than conventional advisory leaflets or booklets.

The most successful warnings are those that contain a single word, such as 'stop', or a short statement that details a hazardous situation, such as 'slippery surface'. Equally effective are those that convey the consequences of ignoring the warning, such as 'ingestion of this product will result in death', and warnings that indicate how the dire consequences can be avoided, such as 'do not allow product to come in contact with hands, face or mouth'. Because warnings tend to be quite brief, they can be positioned close to the area of concern. The position of the warning relative to the area of concern determines the perceived relationship between the two. Warnings such as 'take a break every hour', either located on or directly behind a desk or presented on the screen, achieve such a result. Warnings displayed on the screen can be designed to be intrusive, intending to stop users from continuing with their work and forcing them to attend to the warning.

Some companies have adopted novel approaches to training their workforce. For instance, a method which focusses on younger employees has been developed, which is based on the study of Wang et al. (2014). To reduce the risk of users sitting at their desks for long periods, and to promote healthier computer use, the authors developed a timed broadcast of health-related animations for users sitting at computers for prolonged periods. The animation programme had a positive effect by reminding participants to take a break and stretch their bodies. It was found that overall the programme positively influenced the beliefs and behaviours of participants with regard to their health.

Another approach has involved the use of an innovative self-modelling photo-training method used with the intention of reducing musculoskeletal risk amongst computer users. This was originally developed by Taieb-Maimon et al. (2012). The computer users initially participated in normal group training on how to use their workstations correctly and what postures to adopt. Subsequently, an automatic frequent-feedback webcam system was used. This displayed a photo of the worker's current sitting posture on their computer screen, along with a photo of the 'correct' posture taken of them previously during the standard group training. The use of photography or video as an effective teaching mechanism generally has been well established by a number of studies (e.g. Carney et al. 2010; McGehee et al. 2007). This method has now moved successfully into the world of Ergonomics/Human Factors training (e.g. Jamjumrus and Nanthavanij 2008).

Some businesses employ online training systems that offer an adaptive user interface as the user tries to learn how to use, for example, new equipment. The systems work most effectively when the complexity of the training material is adapted to suit the experience of the user. This is particularly effective when the adaptive training is implemented as part of a product itself, such as when providing users with new tablets. Brudera et al. (2014) found that when the complexity of the training application was adapted according to the learner's experience, learners performed better, were able to undertake new tasks with less help, and had better knowledge about the electronic device.

8.6 Making Training More Effective

Irrespective of the form the training takes, there are a number of aspects that will enhance it. In the first instance, motivation on the part of the trainee is a fundamental part of learning. The amount of time that should be spent in the training sessions should be carefully planned, rather than a trainer arbitrarily filling half a day or a full day. The amount that should be learned in one session should be planned so that the trainees have a target. It should be acknowledged that the trainees do not need an exhaustive account of the subject; they simply need enough information to allow them to work comfortably and safely. Any practical demonstrations should allow the trainees to practice rather than simply watching the trainer going through the motions. The training should take a tactful approach to pointing out any faults displayed by the trainees, and the trainees may need to be reminded that they will have to work at the techniques to get them right. The trainees need a question-and-answer phase and should feel comfortable about raising their concerns or even their disagreement. The end of the session should offer a summary to remind the trainees of the salient points and the reasons behind the training.

After the training has taken place, it is possible for the trainees to return to their normal work routine and completely ignore the advice they have been given. Training can fail for a number of reasons, and these need to be investigated so that future failures can be avoided. Typical reasons for training proving to be a waste of time, money, and effort include inaccurate identification of the type of training that was required in the first instance. If goals have not been set at the beginning of the training phase, the trainees will not know what is expected of them once it has been completed. Training that is perceived to be forcing a new way of working on individuals with a 'you will' approach—rather than an educative approach that 'will enable them' to achieve a change in working behaviour—will be less successful. If trainees attend a course with an attitude that the session simply offers them a break from their work, as opposed to a vehicle to improve the way they work, it is less likely to be successful. Using too much 'chalk and talk' training rather than practical training is not helpful in developing practical skills. Unsuitable, unsuitably trained, or undertrained trainers are unlikely to have the necessary skills to present an effective course or to present the necessary information in the given time frame. Shortened training courses designed to fit into team briefing sessions are unlikely to be long enough to allow trainees to assimilate the quantity of information required to make appropriate changes in the workplace, nor to attempt to practice the skills before returning to the workplace.

The resources made available at the time of the training have an impact on its success. A noisy training environment, close to normal work areas, causes distraction. Poorly functioning equipment interrupts the flow of the material, making it disjointed and distracting. Using dated technology, in terms of both equipment and presentation aids, causes a course to lose credibility, as does using dated statistics or other supposedly supportive material. Training presented in the latest style or with a more 'off-the-wall' approach may not be accepted as readily as material presented in a more straightforward manner. Superiors failing to set the example and taking the approach of 'Do as I say, not as I do' will undermine a course's effectiveness. Courses run as a means of point scoring among different sections within an organisation, or as a means to show where the power lies in terms of having the authority to finance courses, will also lessen the positive impact of a course.

In addition to designing what is thought to be a suitable training programme, businesses should evaluate the impact or effectiveness of the courses. Furnham (2005) believes this is probably the most difficult thing for a company to do. There is no point in churning out the same course time after time if the trainees are uninterested and if they are not subsequently changing their approach to the way they work—which is the ultimate aim of the course. One of the easiest and most immediate ways of evaluating training is to ask participants to complete a feedback or evaluation form at the end of the course. This is a qualitative assessment. An example is shown in Table 8.4. Participants may be more honest when completing this form if they do not have to write their names on it. It should be kept in mind that most evaluation forms are not necessarily a measure of the success of the course in terms of how the attendees will be able to apply their newly acquired knowledge or skills once back at work. Evaluation forms simply record a trainee's personal views on the content of the course, the presenter's approach and ability, and how trainees see the material applying to their own working conditions. Observations after the course of individuals at work is the only way to determine whether the appropriate skills have been learned and applied. However, even this may not give a true picture of what they do when not being observed. Workers often adopt an 'ideal' working posture when they are aware that they are being observed and revert to a more sloppy position once they are not being scrutinised. This is when supervisors and managers should remind them as they walk past about the need to use appropriate work techniques. Wagner et al. (2010) concluded that the combination of both support and training is extremely important, and that good support after training will lead to higher levels of self-efficacy, confidence, a more positive attitude, and reduced anxiety. Brudera et al. (2014) emphasised that, in particular, older people will need support and training if computers, and more specifically mobile devices, are going to be used effectively by everyone.

TABLE 8.4

Example of a Course Evaluation Form

COURSE EVALUATION FORM

We hope that you found the Course both useful and enjoyable. To help us plan future events and to assist in our quality control, we would appreciate it if you could spend a few moments completing the evaluation below.

Thank you for attending and completing the evaluation.

Your name (leave this blank if you prefer):..

Subject of the Seminar/Course: ....

Venue:...Date: ....

1) Assessment of the event as a whole (please tick as appropriate)

	Excellent	Good	Average	Poor
Comprehensiveness?				
Clarity?				
Value to your work?				

2) Standard of Presentation

Please rate in numerical order on a scealof 1 – 5 as follows:

1=Strongly Agree, 2=Agree, 3=Neither Agree or Disagree, 4=Disagree, 5=Strongly Disagree

Were the presentations:	1	2	3	4	5
Relevant?					
Well structured and logical?					
Interesting?					
Professionally presented?					
At a level to suit your needs?					
Were the visual aids:	1	2	3	4	5
Useful?					
Clear and legible?					

Were the discussion periods:

(a) Adequate? Yes ☐ No ☐

(b) Well organised? Yes ☐ No ☐

(c) Well handled by the speakers? Yes ☐ No ☐

Please add below any additional points you would wish to make about the standards of speakers, presentation etc. (please remember we need to know, whether good or bad, if we are to maintain/improve our standards):

....

....

....

(*Continued*)

TABLE 8.4 (*Continued*)

Example of a Course Evaluation Form

3) Finally, please rate the course overall on a scale of 1 to 10 (please tick)

Poor									Excellent
1 ☐	2 ☐	3 ☐	4 ☐	5 ☐	6 ☐	7 ☐	8 ☐	9 ☐	10 ☐

4) Any other comments?

....

....

....

If an evaluation form is not used at the end of the training programme, trainers have to ask themselves a number of questions if they want to develop an understanding of the success of the course. In the first instance, most experienced trainers get a feel for the mood of the group as they present their material by observing their reactions, listening to their questions and comments, and seeing how they approach any practical exercises. The trainer needs to ask: 'Were the course participants content during the course?' If not, they need to probe why not. Perhaps the content of the course did not have any relevance to their working environment. Perhaps the practical aspect of the course did not seem applicable to their environment. For instance, the trainer might have used chairs of a different design from those used by the trainees in their workplace to demonstrate adjustments. Or, the trainees may have been 'volunteered' to go on the course and are unaware of its purpose or may see no value in it. This cannot be a good precursor to a successful course.

The next question trainers need to ask is whether the content of the course actually gets the message across. They can determine this by observing the trainees during any practical exercises. They can observe the length of time it takes the trainees to achieve the desired result and how successful they are at achieving their goal, which might simply be altering the setting of an adjustable chair relative to a desk and keyboard. If the trainer determines that someone's performance in the practical exercises is not satisfactory, they have to consider why. It could be that the way the material was presented verbally was not suitable; the practical exercise might not have been positioned at an appropriate point in the session; or the trainer may conclude that their own training style or knowledge base is not sufficient to run the course successfully.

The trainer then needs to evaluate how well the newly learned material has been transferred into the working environment. This can be done through direct observation of trainees, carrying out discussions with them, or asking them to complete a questionnaire. The timing of the evaluation is important. The greater the time between the end of the training and the start of the evaluation, the less reliable the results. If the trainees are not employing their newly acquired knowledge, the trainer needs to ask whether the course material was not relevant, or whether it was presented in a more complex manner than was necessary thereby confusing the course participants. Alternatively, trainers might have presented the material in such an overly simple manner that it became boring and the trainees stopped attending. The course itself might not have been long enough to pass on all of the necessary information or to allow development of an appropriate skill base.

The last question that the trainer should ask is: 'If the trainees do apply their newly acquired knowledge, will it have a positive impact on their well-being?' If the answer is 'no', trainers need to ask why, and this requires them to re-examine the course content and aims. Once the trainer has validated and evaluated the course, they should make a formal record of the fact that they did so. This enables any future questions regarding the effectiveness of the course to be dealt with informatively. Finally, trainers can carry out a quantifiable assessment in which they measure any changes in levels of absenteeism in a particular area or a change in reports of discomfort among a specific team of workers.

8.7 After the Course

Having established the most effective form that training and education might take within an organisation, plans need to be made for refresher training. It should be assumed that all of the suitable working practices and postures introduced following the original training will start to degrade over time. The situation will be compounded following the employment of new, untrained personnel. Refresher training should occur at least every two years, if annual refresher training is not a viable option.

The organisation's investment in training will only prove worthwhile if mechanisms are put in place to ensure it is adopted and maintained in the day-to-day working environment. Given the possible adverse consequences of adopting an unsuitable posture for extended periods when operating a computer or developing a work routine that results in intensive screen-based work with little or no variety, it is essential that employees adopt and maintain the safe working practices identified during the training and instruction. This is where good management and supervision pays dividends, or at least good leadership does. Leadership is a force that inspires a group of individuals to attempt something different or better. It is an agent of change. This is different from management, which can be defined as a mechanism that focusses on planning, coordinating, and controlling the routine functioning of an office. Those responsible for overseeing the workplace once the trainees return to their normal work should keep in mind that in order to maintain desirable behaviour—and in this context, this equals the adoption of suitable postures and appropriate working practices—they need to provide effective reinforcement and reward schedules. It is quite surprising that few of the individuals who have this responsibility understand how to achieve in their workforce the acquisition of certain behaviour patterns and the extinction of others.

People learn to change their behaviour, such as changing their working posture, so that their employer achieves the desired outcome of risk reduction through positive reinforcement. People's behaviour usually produces outcomes, and, if those outcomes are pleasurable, they will repeat that behaviour. Therefore, if employees are rewarded in some way for changing their working behaviour, they will be more likely to maintain that change in behaviour in the hope that they will be rewarded again. The reward may simply take the form of a manager remarking on how much they have improved their posture. This acts as positive reinforcement. It will be effective only if provided on clear evidence of the desired behaviour so that the employee perceives a clear link between the adoption of the suitable posture and the positive remarks received from the manager. If reinforcement is to work effectively, the rewards should vary with performance: the better the change in behaviour, the better the reward. This will ensure that those who make the most effort do not feel that it was not worthwhile when they realise that those who have made little effort receive the same reward.

Some employees are 'encouraged' to change their working behaviour as a result of experiencing unpleasant outcomes if they maintain undesirable behaviours. What they are actually doing is avoiding undesirable consequences, such as reprimands or even more formal disciplinary action. This process is referred to as 'negative reinforcement'. If a certain behaviour results in the cessation of an unpleasant event, the individual is likely to repeat it. Punishment is different from negative reinforcement and may not be as effective at eliminating unwanted behaviour. If individuals are punished for working in a particular way, for instance, by being heavily criticised by a manager, they will simply associate the undesirable outcome with the unwanted behaviour. They will not necessarily see outside of this immediate combination to the connection between the desired behaviour and the removal of the undesirable outcome. If punishment is to be used, it works most effectively if delivered immediately after the undesirable behaviour is witnessed, as this enables the individual to connect the two events. The longer the delay between the undesirable behaviour and the punishment, the weaker the association between the two, so it becomes less likely that the undesirable behaviour will be suppressed. Allowing the undesirable behaviour to persist for some time without punishment can lead to strong reinforcement of it, making it more difficult to change. For instance, if someone chooses to sit with his chair too low for his desk and keyboard, and a manager walks past every day without making a comment, this sends a message that this behaviour must be acceptable. In this instance, failing to react to the individual's way of sitting when at work has reinforcing consequences. The punishment should be moderate, as opposed to mild or severe. If it is mild, the employee may ignore it or become accustomed to it.

If it is severe, such as dismissing the individual for failing to adjust the chair properly, they may reject the punishment as unfair and out of scale to their misdemeanour. Although remaining within 'moderate' bounds, the punishment should be increased progressively. In other words, as the behaviour is repeated following each reprimand, the manager may have to become more critical, then may have to move on to a written warning, and so on. The punishment should be impersonal and should be focussed on the undesirable action, not the individual. For instance, stating that someone is working in a poor posture because they are too stupid to work out how to operate their chair is a personal attack that will humiliate them. Punishment should be used consistently with all employees and should be used for every instance of unsuitable behaviour identified.

Sometimes two employees may be displaying the same inappropriate working behaviour, but the circumstances differ, leading to a different interpretation of the events and different treatment of the individuals. For instance, two people may locate their monitors to the side of their desks causing them to twist when working. Investigation might reveal that one is working in that manner because 'that is what I have always done', while the other may be doing it because the cable on the monitor is too short to allow it to be moved away from the power point towards the centre of the desk. Maintaining consistency means it should be maintained by all managers in the same way. Employees must understand exactly why they are being punished, and, therefore, managers must be able to communicate their reasons for criticising someone's work routine. This would normally be done in a disciplinary meeting in the manager's office. During this meeting, the manager should refer to previous problems regarding an employee's behaviours and ask why the employee was unable to change the manner in which they work. This should be followed by a verbal warning, which needs to be recorded in the personnel file. There is a suggestion that giving responsibility to a team or group of individuals to monitor one another—in other words, developing an autonomous group—can be beneficial in ensuring that suitable working practices are employed.

It should be kept in mind that some individuals will learn new behaviours vicariously by simply observing what attention is focussed on the others around them. Therefore, changes in their behaviour may occur as a result of watching others being praised or criticised.

If an organisation wants to maintain the desired behaviour once the training has been completed, it should develop a specific programme designed to achieve this outcome. In the first instance, it should specify the desired behaviour in clear terms. For example, instead of stating that employees are required to develop more suitable working practices, the organisation should spell out that it expects that workstation furniture will be arranged and used correctly, utilising the guidance provided in the training, and that employees should manage their work to incorporate regular, albeit brief, breaks in their screen-based work. The organisation should have a baseline profile of where its employees stand currently, so that it can identify future improvements. It should then define the standard expected of the workforce, and, following that, it should decide on how to reinforce that standard. This needs to be something that can be delivered easily and promptly when required. After this point, the company can 'shape' behaviour by reinforcing behaviour that is close to what is expected, although not perfect. The reinforcement becomes more frequent as the individual's behaviour gets closer to the required standard. This whole process should be reviewed at intervals to ensure that the expected behaviour is still evident, and that the rewards are continuing to have the desired impact. Companies should not be disheartened if changes to the desired behaviour have occurred as this is to be expected.

In summary, the most effective training programmes are based on three basic principles: 'Plan, 'Do', and 'Review'. That is, plan how the behaviours of the workforce can be improved; implement action that will ensure that education is offered that enables the changes to be made; and evaluate the outcomes to determine if the desired goals have been achieved. To be meaningful, the evaluation has to measure exactly what has improved and in what ways.

8.8 Summary

- Training is the means by which an individual acquires a set of rules, concepts, skills, and attitudes.
- Training should be part of an overall programme to develop a positive attitude to health, safety, and well-being in the workplace.
- Training should not be viewed as a substitute for proper risk control.
- A company's view of training will influence its success.
- Positive reinforcement influences behaviour.
- The value of feedback is influenced by who dispenses it.
- Frequent short breaks during training result in rapid acquisition of skills.
- If many people need to be trained, they should be prioritised so that those who need it most receive it first.
- Supervisors and managers should receive training similar to that given to the rest of the workforce so they understand what trainers are trying to achieve.
- Trainers should know in advance of a course whether any trainee has special needs so they can accommodate them.

- Trainees should sign an attendance sheet which should be stored for future reference.
- Personal development plans allow employees to define their own personal training needs, which allows the training to be tailored to suit their needs.
- The success of the course will be influenced by the skill of the trainer.
- Trainers should have good presentational skills, generate a rapport with their trainees, be able to deal with difficult or argumentative individuals, be able to deal with learning at different levels, be capable of designing an interesting and effective course, be able to evaluate trainees, offer feedback, and evaluate the effectiveness of their own course.
- Both in-house and external trainers have their own advantages and disadvantages.
- Trainers are considered to be persuasive by a group if they are perceived to be credible and attractive. Credibility is influenced by competence when training, and attractiveness is influenced by personal factors such as personality and appearance.
- The content of the course should be driven by what the trainees need to achieve at the end.
- Risk assessments are a good means of identifying what training is required.
- The nature of the material and the way it is delivered will have an impact on the effectiveness of training.
- Vivid imagery is more effective than bland statistics.
- Trainees have a preferred learning style, and if this fits the trainer's teaching style the training will be more successful.
- Training in workstation use and posture should include a practical session.
- Training can be done in a traditional lecture format, one-to-one, by watching video clips, and through online training. Each has its advantages and disadvantages.
- Videos clips are helpful but do not guarantee a transfer of information to the workplace.
- Alternative effective training methods are available, such as systems that use webcams, on-screen animated characters, and displayed before-and-after photographs.
- One-to-one coaching allows the trainer to ensure that each workstation is used correctly, and allows each individual to ask questions.

- Online training does not allow for questions to be asked and does not guarantee absorption of the information.
- Leaflets are not an effective means of educating people.
- Warnings displayed at the workstation can be effective if single words or short phrases are used.
- Training should be followed up to ensure that good practices are maintained once the trainees are back at work.
- Trainees should know what is expected of them upon completion of the course.
- The resources made available will have an impact on the success of the course.
- Training programmes should be evaluated to determine their effectiveness and altered if they are not working. Questions need to be asked about why the course was not effective so appropriate changes can be made.
- Supervisors and managers should lead by example.
- Good leadership should inspire people to try something different.
- Positive reinforcement is an effective means of changing behaviour.
- Some organisations use negative reinforcement and punishment as a means to change behaviour.
- Some people learn new behaviours vicariously by observing other's behaviour and seeing whether they are praised or criticised.

References

Brudera, C., Blessing, L., and Wandke, H. 2014. Adaptive training interfaces for less-experienced, elderly users of electronic devices. *Behaviour & Information Technology* 33(1), 4–15.

Carney, C., McGehee, D. V., Lee, J. D., Reyes, M. L., and Raby, M. 2010. Using an event triggered video intervention system to expand the supervised learning of newly licensed adolescent drivers. *American Journal of Public Health* 100(6), 1101–1106.

Furnham, A. 2005. *The Psychology of Behaviour at Work: The Individual in the Organization*. 2nd ed. Hove, UK: Psychology Press.

Jamjumrus, N., and Nanthavanij, S. 2008. Ergonomic intervention for improving work postures during notebook computer operation. *Journal of Human Ergology* 37, 23–33.

Kroemer, K. H. E., and Grandjean, E. 1997. *Fitting the Task to the Human: A Textbook of Occupational Ergonomics*. 5th ed. London: Taylor and Francis.

Lambert, T. 1993. *Key Management Tools*. London, UK: Pitman.

Loh, Peggy Y. W., Lo, M. C., Wang, Y. C., and Mohd-Nor, R. 2013. Improving the level of competencies for small and medium enterprises in Malaysia through enhancing the effectiveness of E-training: A conceptual paper. *Labuan e-Journal of Muamalat and Society* 7, 1–16.

McGehee, D. V., Raby, M., Carney, C., Lee, J. D., and Reyes, M. L. 2007. Extending parental mentoring using an event-triggered video intervention in rural teen drivers. *Journal of Safety Research* 38(2), 215–227.

Taieb-Maimon, M., Cwikel, J., Shapira, B., and Orenstein, I. 2012. The effectiveness of a training method using self-modeling webcam photos for reducing musculoskeletal risk among office workers using computers. *Applied Ergonomics* 43, 376–385.

Trichesa, R. M., and Giuglianib, E. R. J. 2005. Obesity, eating habits and nutritional knowledge among school children. *Revista de Saúde Pública* 39(4), 1–7.

Visschers, V. H. M., Ruiter, R. A. C., Kools, M., and Meertens, R. M. 2004. The effects of warnings and an educational brochure on computer working posture: A test of the C-HIP model in the context of RSI-relevant behavior. *Ergonomics* 47(14), 1484–1498.

Wagner, N., Hassanein, K., and Head, M. 2010. Computer use by older adults: A multi-disciplinary review. *Computers in Human Behavior* 26(5), 870–882.

Wang, C., Jianga, C., and Chern, J.-Y. 2014. Promoting healthy computer use: Timing-informed computer health animations for prolonged sitting computer users. *Behaviour & Information Technology* 33(3), 295–301.

Zainab, B., Bhatti, M. A., and Alshagawi, M. 2017. Factors affecting e-training adoption: An examination of perceived cost, computer self-efficacy and the technology acceptance model. *Behaviour & Information Technology* 36(12), 1261–1273.

9

The Environment

9.1 Introduction

This chapter is not intended to offer an in-depth analysis of all environmental factors in an office. Such an appraisal should be sought in a publication dedicated solely to this subject. Instead, this chapter offers an insight into factors that are common causes of complaint amongst the workforce. Ergonomists and Human Factors Consultants are typically concerned with the impact of three main factors in an office environment: noise, lighting, and thermal comfort. In an office that incorporates any heavy machinery, such as printing equipment or mail-sorting machines, vibration is also likely to be of interest.

This chapter offers some general comment on the three environmental factors, as well as some advice on how to create a suitable office for people to work in comfortably. It is aimed at traditional office buildings that have environments that are unlikely to change from day to day or hour to hour, as might occur when using mobile devices on the move.

9.2 Noise

Acoustic waves can be described as fluctuations in pressure, or oscillations, in an elastic medium. The oscillations produce an auditory experience which is sound. This is achieved because the ear converts the acoustic waves into nerve impulses, which move to the brain along the auditory nerve. The brain processes this information and imposes some sense on it, resulting in perception of sound and identification of auditory patterns. How loud a sound is considered to be is determined by its frequency and its sound pressure level (SPL). Frequency refers to the complete number of cycles that occur in one second. It is expressed in Hertz (Hz) and gives the sensation of pitch. The amplitude of the sound wave corresponds to the intensity of the sound and provides the sensation of loudness. The human ear is normally sensitive to a range of frequencies between 20 and 20,000 Hz; this is referred to as the audible spectrum. We are likely to hear at our best between 1000 and 4000 Hz, which is the frequency band in which speech is transmitted. Auditory thresholds—the point at which we can actually hear something—are much lower at higher frequencies. Noises at lower frequencies have to be much louder to be heard. Noise is measured in decibels (dB). An A-weighting, written as dB(A), is used to measure average noise levels. A C-weighting, written as dB(C), measures peak, impact, or explosive noise. Table 9.1 indicates typical noise levels encountered in a number of situations.

TABLE 9.1

Typical Noise Levels Encountered in a Number of Situations

dB	Situation
160	Explosion
150	Rupture of ear drum
140	Jet engine at 25-30 metres
130	Pneumatic chipper
120	Riveter
115	Punch Press
110	Chainsaw Powered lawnmower
100	Jet flyover (min 1000 feet) Night club bar
95	Power drill Arc welder
90	Motorcycle (7metres) Personal stereo (using ear pieces)
85	Diesel lorry Tractor cab
80	Lathe Heavy traffic
75	Primary school classroom
70	Television Radio
60	Noisy office Conversation
50	Typical office
40	'Quiet' office
30	Quiet in the countryside Library
20	Whisper one metre from ear
10	Soundproof room
0	Threshold of hearing

Noise is defined as unwanted sound. It is a subjective experience, whereas sound can be measured objectively. Sound does not have to be loud to be noise; it simply needs to be annoying, distracting, or unacceptable to an individual. In an office setting, noise can interfere with people's level of concentration and can affect their performance. It can make communication with other people in the office, or on the telephone, more difficult. This is a typical complaint in open-plan offices. The real conflict in an open-plan office is that people are placed in these environments to enhance communication between them (among other things), but they still need a degree of privacy to work effectively. It is difficult for a single environment to accomplish all of this successfully. When compared to enclosed individual offices, the open-plan environment produces reduced satisfaction and performance. A reduction in performance is reflected in deterioration in concentration, memory, and learning. It is believed that one of the main reasons for this occurring is reduced 'auditory privacy' (de Croon et al. 2005; Jahncke and Halin 2012; Jahncke et al. 2013; Kaarlela-Tuomaala et al. 2009; Kim and de Dear 2013).

Because of the disruptive influence of noise in open-plan offices, employers are turning to activity-based offices. Activity-based offices commonly involve open-plan settings (Wohlers and Hertel 2016), however, the aim of the activity-based offices is to provide a diversity of settings which may ameliorate the dissatisfaction and reduction in performance resulting from a lack of auditory privacy by allowing people to move to a quiet setting. Seddigh et al. (2014) found that there are fewer distractions in activity-based offices compared to open plan environments. A clear difference between the two environments is office use: open-plan offices have assigned workstations, while the activity-based office applies a non-territorial workplace concept with flexi-desking or hot-desking (Rolfo et al. 2015, 2018).

As an alternative, Helenius et al. (2007) has suggest that increasing efforts should focus on the acoustic design of open-plan offices rather than change the concept of how the office functions. For instance, Passero and Zannin (2012) found that the insertion of dividers between work stations and an increase in the ceiling's sound absorption improved the acoustic conditions in the office.

Persistent noise can result in negative emotions, such as anger and frustration for employees. Their subjective response appears to be determined more by the nature and context of the noise than by its level or intensity. For instance, hearing colleagues chatting at adjacent workstations tends to be more disruptive than the noise from the air conditioning unit above the desks. The difference is that the air conditioner produces a consistent, unstructured noise to which people become habituated. The overall background noise level within an office determines whether specific intermittent noises are distracting. For instance, a telephone ringing in a noisy office will probably not be noticed as much by the workforce as a telephone ringing in a quiet environment like a library. Moreover, different people will respond differently to noises, with some being more tolerant than others.

Noise has other characteristics which will influence how annoying it is perceived to be in any workplace. If the noise is variable, where the loudness or frequency changes, this is considered to have a high level of annoyance. The sound quality will cause high levels of annoyance if it is considered to be high pitched or deep and rumbling. The information content will prove to be highly annoying if it is meaningful, such as a conversation. Even low-level background speech of high intelligibility significantly impairs short-term memory, reasoning ability, and well-being (Liebl et al. 2012). If the noise is considered to be within the control of another individual in the office/building, it will create a higher level of annoyance. Alternatively, if the individual feels they have more control over the level of noise, they can predict its occurrence or they view the noise as being an unavoidable and necessary part of their work or working environment, they will experience a lower level of annoyance. This latter point is considered by some to display resignation on the part of the individual.

It is generally accepted that noise levels that do not exceed 85 dB(A) are unlikely to cause harm to an individual's hearing. However, it is generally recommended that full risk assessments be carried out if workers are exposed to 80 dB(A) for more than six hours a day. Repeated exposure to noise levels in excess of 85 dB(A) during an extended period of employment can result in permanent hearing loss. Noise levels well below 85 dB(A) can be annoying and distracting, which can affect performance. Table 9.2 outlines a summary of findings by Woodson (1981) reflecting the deterioration in performance that can occur as a result of different noise levels. More specifically, in relation to having a meeting or discussion in the office, Table 9.3 shows the impact of increasing noise on the quality of a two-way conversation in terms of being heard and understood.

The noise levels within an office depend not only on the loudness of a sound, but also on the reverberation potential within that area. The reverberation level affects people's perception of how noisy they find their office. Reverberation relates to the 'bouncing' of sound waves off surfaces, such as walls. Reverberation time gives an indication of how long it takes for the reverberant level to fall by 60 dB, and this is influenced by how large the room is and the amount of sound-absorbing surfaces within it. The fewer the sound-absorbing surfaces, or objects within a room, the greater the reverberation. It is not influenced by where the listener stands or where the source of the sound is located. High levels of reverberation interfere with speech perception. Carpeting and partitioning can act as sound absorbers.

There will be times in all offices when people need to communicate with each other regarding their work. Listening and understanding a target sound, such as occurs when discussing a project with a colleague at an adjacent desk, can be adversely affected by background noise. The adverse effects are increased if the target sound and the distracting sound share similar frequencies. In other words, it is more difficult to listen to and understand an individual at an adjacent desk if other people are talking in the immediate vicinity. This type of disruption is known as 'masking'. It has been suggested

TABLE 9.2

Table Showing the Deterioration in Performance as a Result of Noise

<30	Low-level intermittent sounds becoming disturbing.
40	Very acceptable for concentration. Few people will have sleep problems.
50	Acceptance by people who expect quiet. About a quarter of people will experience difficulty in falling asleep or be woken.
55	Upper acceptance level when people expect quiet.
60	Acceptable level for daytime living conditions.
65	Upper acceptance level when people expect a noisy environment.
70	Upper level for normal conversation. Telephone conversation difficult. Unsuitable for office work.
75	Telephone conversation difficult. Raised voice needed for face-to-face communication.
80	Conversation difficult.

Source: Woodson, W., *Human Factors Design Handbook*, McGraw-Hill, New York, 1981.

TABLE 9.3

Changes in Extent of Disturbance as Noise Levels Changes

Noise Level (dB)	Quality of Discussion
<40	Excellent
40–45	Very good
45–50	Good
50–55	Acceptable
55–60	Impaired
65–80	Hindered
>80	Severely restricted

that masking of this type might interfere with short-term memory. Because short-term memory uses verbal coding to an extent, surrounding noise can mask the 'inner voice' used to code the information, resulting in a degradation in performance. This type of problem is experienced by many individuals, who would describe it as 'not being able to hear myself think' when working in a noisy environment.

Although it is usually accepted that exposure to loud noise can damage hearing, studies are now suggesting that it can have a far greater effect on general health than might have been realised. Consistent exposure to noise levels in excess of 89 dB(A) has been related to increased blood pressure. In less extreme environments, a reduction in motivation has been recorded in open-plan offices where the noise levels were well below what might be considered 'noticeable' (Evans and Johnsson 2000). This latter point indicates that even noise levels that fall within what might be considered acceptable are still likely to have a negative impact on the workforce and should be monitored.

Employers need to control the noise levels in their workplaces in order to maintain performance at acceptable levels and reduce workers annoyance. Reducing noise to a level where it is inaudible would be impractical and expensive. However, methods can be employed to reduce the extent of the disruption from sources of noise. For instance, unanswered telephones should be diverted in the absence of individuals from their desks. Equipment can be made quieter through the use of aids and attachments available from many manufacturers. Those needing to concentrate should be positioned away from noisy walkways or work areas. Whatever steps are taken, however, an employer should be prepared to achieve the opposite of what was intended. It has been suggested (Banbury and Berry 2005) that attempts to reduce noise in an environment may exacerbate the problem rather than alleviate it, because in quieter offices, noise is more discernible or 'noticeable' and the inhabitants less tolerant of it because it is more distinctive. In a noisier office, there is more likely to be a masking effect as each noise becomes less distinct or discernible relative to another. Some researchers have even suggested adding noise to an environment to make it less disruptive, because individual sources of noise become less distinct as they are masked by other sounds. This is referred to as 'noise perfuming'. Banbury and Berry (2005) have concluded that the level of white noise necessary to mask disruptive noise is impractical for use in offices.

As offices do not remain in a status quo, but tend to change weekly, monthly, or annually, employers should appreciate how much the noise levels need to change before they will be noticed by the workforce. Changes might occur as a result of the introduction of new equipment, such as the installation of new telephone lines or a new printer. It appears that increasing or decreasing noise levels in an office environment by 5 dB is sufficient to be noticed (in contrast to 1 dB under laboratory conditions). The more dominant the new noise source is in an environment, the more noticeable its introduction.

9.3 Lighting

Lighting in offices is not a straightforward subject to tackle. Before making any attempt to identify whether the lighting is suitable for the workers, an understanding of the work they perform is required. A clear distinction needs to be made between workers who perform screen-based tasks and those who perform paper-based tasks since they will require two different levels of lighting in order to perform their work satisfactorily. It also needs to be understood that there is likely to be a difference between creating a lighting atmosphere within an office that suits the work performed there and creating a lighting atmosphere that people will like. It is not always possible to achieve both, although that should be the ultimate aim of any lighting system.

When the subject of lighting is discussed, a number of terms are commonly used and these include luminance, illuminance, luminous intensity, and luminous flux. Luminance refers to the light emitted, or bounced back, by a surface toward the eye. This is measured in cd/m^2, cd being an abbreviation for candela. Illuminance relates to the amount of light falling on a surface, usually horizontal surfaces, and this is measured in lux. High levels of illuminance may cause glare and result in 'washing out' of items being viewed. Reflectance is an indicator of the ratio of luminance to illuminance. When an organisation specifies the level of illuminance required in an office, it should also specify the corresponding reflectance of surfaces in the office, as this is more likely to result in a balance of surface luminance. Luminous intensity, which is measured in candela, relates to the amount of light emitted from a source, and luminous flux indicates the flow rate of the luminous energy. The luminance distribution within the visual field should ensure that it is well-balanced, tasks can be completed unimpaired, glare is avoided, visual communication is heightened, and safety is uncompromised.

The luminance surrounding a computer display can potentially reduce visibility of the display (disability glare), result in sensations of discomfort (discomfort glare), and result in transient adaptation effects from fixating back and forth between the two luminance levels (Sheedy et al. 2005).

9.3.1 Glare

Glare occurs when there is an imbalance of surface luminance within the visual field. Glare can be either direct or indirect—that is, emitted by a source or reflected off a surface (see Figure 9.1). Indirect glare can be subdivided into diffuse, such as a light-coloured wall reflecting light, or specular, a mirror-like glare typical when reflected off a glass panel or highly polished metal surface. Glare should be avoided through good design and thoughtful installation of light sources.

Although the retina of the eye is able to adapt to different levels of luminance so that people can function effectively as they experience different environments, it encounters difficulties when presented with single, large disparities in the visual field, such as a very bright desk light used in a dimly lit office. Direct glare can be caused by daylight if the individual has a direct view through a window to the sun or even the clouds. This can be controlled through the use of blinds or curtains. However, whatever window covers are introduced, they should not alter the perceived colour climate of the office environment. Their luminance level should also be monitored so that they do not produce unacceptable levels of luminance, for instance, on a sunny day.

The area around the window should have a high reflectance value to reduce the glare effect of a bright window presented in the middle of a dark surround. Direct glare can also be caused by artificial light sources.

Disability glare is considered to make the completion of a task more difficult and demanding and is found in situations where a source of light brighter than that on the actual task causes interference, so that material related to the task is difficult to decipher or interpret. This is different from discomfort glare, where a bright object is identified on the periphery of the visual field, but does not interfere with the completion of a task. To avoid glare, light sources should not be 'seen' within the visual field. For this reason, it is always recommended that office workers should sit between rows of ceiling lights, as opposed to in line with a row of lights, whenever possible. This will reduce the likelihood of them experiencing both direct and indirect glare, as illustrated in Figure 9.1. Workers performing screen-based tasks are more likely to encounter glare problems than are clerical workers carrying out paper-based tasks, because computer users work with their heads at a higher level and their gaze is not directed downwards, as is the case with the paper-based task (see Figure 9.2). Because of this difference between types of workers, screen-based users need more complex consideration.

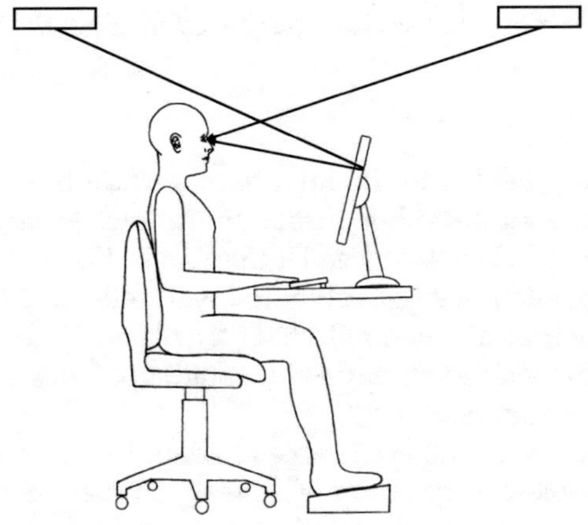

FIGURE 9.1
Direct glare and indirect glare.

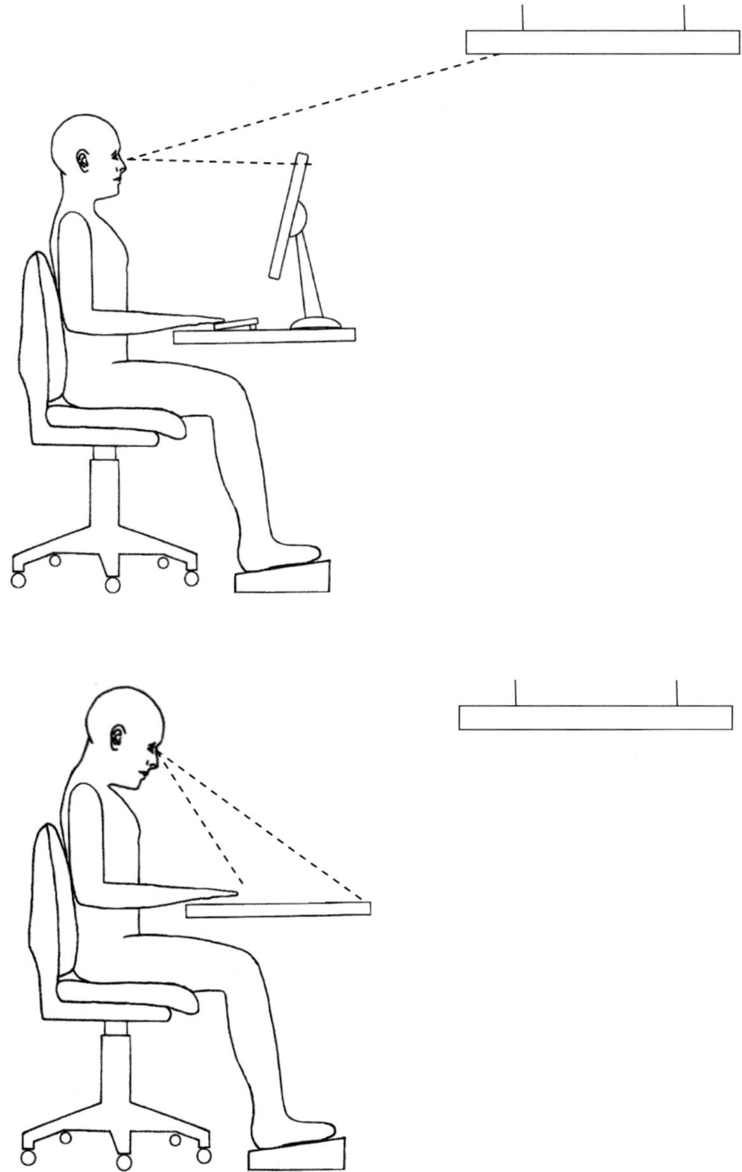

FIGURE 9.2
Why screen users are likely to encounter problems with glare.

9.3.2 Artificial Light

Direct lighting, which is typical of many offices and where most of the light is directed downward in an inverted cone shape, often results in the light being reflected off surfaces, such as desks and documents, causing indirect glare and hard shadows. It can also cause sharp contrasts between illuminated areas and areas that are not illuminated. The most common form of direct lighting is fluorescent lights. These are known to flicker, but the flicker frequency of correctly functioning fluorescent lights should not be perceptible to office workers. Older or malfunctioning lights may have a noticeable flicker, which can be distracting and can cause visual discomfort and headaches in some individuals.

Indirect lighting, where the light is directed mainly toward the ceiling and walls before being reflected back, results in a more consistent distribution of light across the work area and a reduction in shadows. Using wood or materials with a matte finish is more suited to the construction of desks than highly polished finishes. Using a greater number of lower-powered sources of light rather than a few high-powered sources will also reduce glare problems.

Küller et al. (2006) found that the mood of office workers was at its lowest when the lighting was experienced as much too dark. Their mood improved and reached its peak when the lighting level was perceived to be just right. It declined again once the office workers considered it too bright. It appears that there are four characteristics of lighting that are important in terms of an individual's subjective impressions: (1) the light's location, such as overhead or on the periphery, (2) whether it is considered bright or dim, (3) whether the light appears uniform across the work area, and (4) whether it is considered to give visually 'cool' or 'warm' light. It seems that office workers are generally satisfied with an illumination level of about 400 lux. However, suitability of lighting levels is task-dependent. For people performing a screen-based task, lighting levels around 500 lux are probably optimal, as this still permits workers to perform non-screen-based tasks. It would not be bright enough for a paper-based task to be completed with ease. An illuminance level of about 700 lux is likely to be suitable for the latter task. However, working with illuminance levels above this when performing a paper-based task should be acceptable as long as it does not exceed 1000 lux. At this point, office workers are likely to complain about glare, reflections, shadows, and other visual problems.

Providing individual workstation lighting, as well as the standard uniform room lighting, is an effective means of providing lighting that suits the people and the tasks performed at individual workstations. Desk lighting permits the individual to alter the direction and level of lighting. The level will change as the individual's task requirements change. The individual also has control over setting the lighting level to suit their own personal preferences and visual abilities.

People with SAD (seasonal affective disorder) are being treated successfully with phototherapy. The basis of the treatment is to expose them to bright light of up to 2500 lux first thing in the morning and at the end of the day so as to extend the length of their exposure to daylight conditions. On a separate, but related matter, it has been suggested (Laurance 2006) that the 'post-lunch dip' might be influenced by lighting. It is thought that the sensitivity of newly discovered receptors in the eye peaks with light toward the blue end of the spectrum. It has been suggested that these receptors respond to light in a non-visual way, sending signals to the hypothalamus in the brain, which regulates circadian rhythms. Research results have shown that office staff felt more alert and were working more effectively after the lights were changed to a cooler, bluish hue, although they still looked like the standard fluorescent type.

If it is determined after an assessment that the lighting levels are insufficient to allow tasks to be completed without difficulty, a number of solutions need to be considered. These are not necessarily time-consuming or expensive. In many instances, simply cleaning the lamps and/or their covers and shields and replacing failed light sources will be sufficient to make lighting levels acceptable. These steps are frequently neglected in offices. Changing the room surfaces from dark to lighter colours to increase their reflective value and removing obstructions from around light sources will have a positive impact. Moving light sources closer together and adding additional units will be beneficial, or alternatively, moving the workstations closer to the light sources would work equally well. If the assessment shows that the work area is not uniformly lit, the first area for consideration is whether any of the light sources have failed or are dirty. There may be large gaps between light fittings and reducing the spacing or providing additional lighting should provide a more uniform distribution of light. The lights could be altered to give a wider distribution of light or to direct the light upward more. Altering the surfaces within the work area from dark to lighter finishes and removing obstructions will have a similar effect. If the lights are considered to be too bright, they need to be fitted with a control mechanism; alternatively, they could be moved so that they do not fall within the gaze of the workers, causing a glare problem. Another strategy is to reposition the workstations relative to the lights. If the lights cannot be moved, they can be reoriented if they are linear, such as a fluorescent light, so that they are viewed end-on rather than side-on, which tends to produce greater glare problems. The light sources could also be raised, as long as this does not result in an unacceptable drop in lighting levels.

9.3.3 Natural Light

It is generally accepted that people prefer to work in environments with windows, and that lack of windows and a view to the outside may have a negative impact on well-being and performance. It has been suggested that all windows should receive direct light from the sky, and workers should be able to see the sky from their windows. The location of balconies, canopies, and other overhanging structures will have an impact on this. It has been suggested that windows should cover at least 20% of the window wall area if an employer wants most of the workforce to be satisfied by the availability of windows (Rea 2000). Tall windows, rather than broad windows, are considered more effective in providing light, because the light is able to penetrate further into the room. Architects who aim to construct buildings so that externally they resemble glass buildings should appreciate that this causes large problems for the workforce in terms of regulating glare and reflections, as well as extreme heat in summer.

Placing coverings on the windows and using materials to tint the glass are likely to have an impact on mood. Frosting glass can reduce transparency by 30% to 70%. The same applies to glass bricks and glass with insulating properties.

Computer users should, ideally, sit parallel to windows. They should avoid sitting facing a window, which might result in direct glare, and they should avoid sitting with their backs to a window, which can result in problems with reflections on their screens. High-luminance reflections that virtually obscure the detail displayed on the screen are called 'veiling reflections'. Figure 9.3 shows an ideal office layout in terms of overhead lights and windows.

9.3.4 Colour

Küller et al. (2006) showed that good colour design contributed to a positive mood among the workforce. They showed that the emotional status of individuals with a colourful work environment was higher throughout the year than was the case for those who did not work in such an environment. The brighter the colour, the happier the individuals who worked in that office. However, this does not advocate the introduction of strong colours, as previous studies have shown that the use of very strong colours can have a detrimental effect on mood and performance. The findings of Küller et al. (2006) simply suggest that good colour design has a positive impact on the mood of the workforce. It should be kept in mind that the perceived colour of an object is influenced by the type of artificial light used to illuminate the work area. For instance, standard fluorescent lights strengthen the appearance of blue and green objects more than that of red, pink, and orange objects. This consideration has significance if an individual is performing a colour-discrimination task.

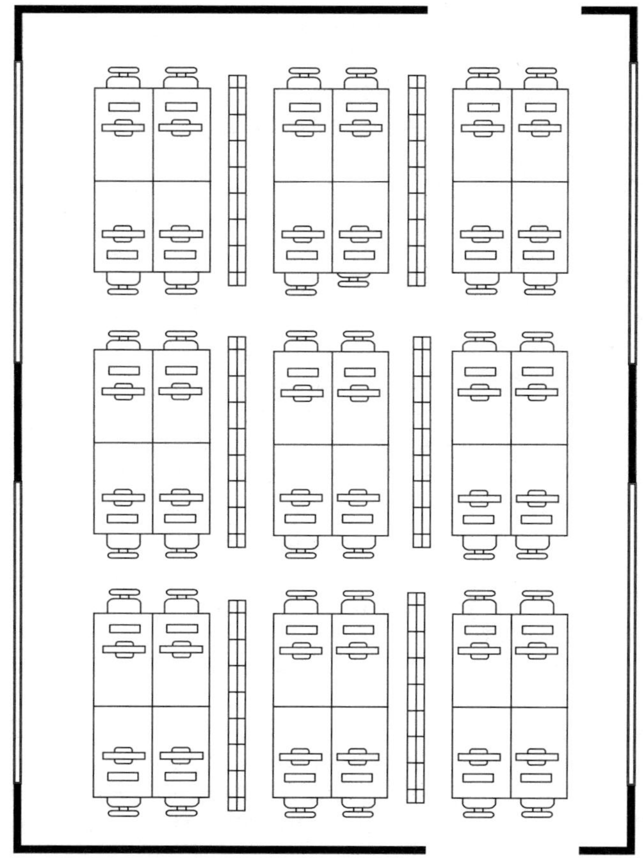

FIGURE 9.3
An ideal office layout in relation to overhead lights and windows.

9.4 Thermal Comfort

Thermal comfort involves the consideration of more than just the temperature of an environment. It also includes an examination of humidity and air velocity (or draughts). It is particularly difficult to get the thermal environment just right for everyone in an office, owing to the fact that the number of people present changes from day to day, they each wear different clothes and the amount of equipment in use changes over time.

Thermal comfort is not something that can simply be measured objectively using equipment, with an assumption being made that if the readings hit the right levels, the environment is suitable. Thermal comfort is not that straightforward, because it is a psychological phenomenon and subjective appraisals need to be included in any assessment of it. This form of assessment evaluates the affectivity of the thermal environment and covers issues such as comfort, preference, and acceptability. Ratings scales are usually used to collect this type of information. The other two criteria that should be evaluated are physiological aspects, such as health, and performance. Table 9.4 provides an example of a thermal comfort evaluation form that also collects information relating to health and performance.

Dealing with the thermal environment is made more difficult, because most purpose-built offices are sealed units that do not allow people to open the windows to change the temperature in any way. Other aspects of building design also influence the indoor climate. For instance, the amount of insulation present affects the heat exchange between the inside of the building and the outside. The same can be said of the type of materials used. It is not only a case of losing heat in the winter; it is also a case of overheating in the hotter months. Heavyweight materials act better as buffers between the outdoor and indoor climates.

Humans react to the conditions within their environment to preserve their core temperature. For that reason, they are referred to as 'homeotherms'. Their physiological reaction to the thermal environment is crucial because failure by the body to respond properly or to be able to cope with extremes can result in ill health. A steep increase or decrease in body temperature can lead to confusion and collapse, and possibly death, in extreme situations. The mechanism the body has to control its core temperature is found in the brainstem and acts like a thermostat. If the environment is cold, blood circulation to the skin is reduced by vasoconstriction to prevent heat loss, which can, not unexpectedly, lead to a drop in skin temperature and an increase in complaints about being cold. It can also cause a reduction in performance. If the environment is too hot, blood flow to the skin is increased through blood vessel expansion (vasodilation), allowing the heat to dissipate. Therefore, blood flow is the most important controlling mechanism in regulating body temperature. Sweating is the next most important mechanism for regulating body temperature, and this occurs if vasodilation has not been effective enough. The third mechanism for regulating temperature raises the rate at which heat is generated by the body, and it does this through shivering. This occurs when the body is excessively cold.

Because the body is constantly regulating its temperature, there is a continual exchange of heat between it and its surroundings through conduction,

TABLE 9.4

Example of a Thermal Comfort Evaluation Form

Evaluation of Thermal Comfort

Name:		**Job title:**	
Building:		**Shift length:**	
Floor:		**Equipment used by you:**	
Department:		**Equipment you share:**	

This evaluation is intended to record your views about your thermal environment. It will also record any related health problems you might be experiencing and the time of day these occur.

Part 1: When at work how often do you feel it is:

	Always	Often	Sometimes	Rarely	Never
Hot					
Warm					
Cool					
Cold					
Dry					
Humid					
Draughty					
Stuffy					
Perfect conditions					

Part 2: How often do you experience any of the following symptoms:

	Always		Often		Sometimes		Rarely		Never	
	am	pm	am	pm	am	pm	am	pm	am	pm
Dry throat										
Sore throat										
Dry skin										
Dry lips										
Dry mouth										
Brittle nails										
Rashes										
Irritation in eyes										
Irritation from lenses										
Runny nose										
Blocked nose										
Breathing difficulties										
Chest discomfort										
Headaches										
Light headedness										
Nausea										
Tiredness										
Lethargy										
Hard to concentrate										
Feeling depressed										

convection, evaporation, and radiation. Conduction occurs when the skin comes in contact with a surface. This is significant in the office, because it has implications with regard to the surface finishes selected for office furniture and equipment. If inadvertent heat loss from the body is to be avoided, office workers should not have to work with surfaces constructed from materials that conduct the heat away from the body, such as aluminum or other metal finishes. This also relates to the edging on wooden desks against which people might lean as they work. If workers have been given cold surfaces to use, these can be covered in materials that reduce conductivity. If the heat exchange occurs through the surrounding air, it is referred to as convection. This is influenced by the temperature of the surrounding air and air velocity. This is normally responsible for about 30% of the heat exchange that occurs. Convection also occurs through air movement, which replaces humid layers of air from next to the skin, increasing the acceleration in water vapour pressure and temperature, and thus cooling the skin.

Even when people feel that they are working in a pleasant environment, they still continue to sweat. Sweating occurs continually throughout a day. If the environment becomes too hot, the skin responds with increased sweating accompanied by a rapid increase in heat loss from the body through evaporation. The extent of heat loss is influenced by the amount of exposed skin from which the sweat can evaporate. The type of clothing being worn determines, to an extent, how effective sweating will be; non-permeable clothing will result in sweat dripping off rather than evaporating, providing little or no cooling. The amount of heat loss is also influenced by relative humidity, which relates to the variation in water vapour pressure adjacent to the skin and further away. If the environment is warm enough to cause an increase in sweating, there is likely to be a drop in the individual's level of arousal.

People radiate heat, and this is exchanged on a two-way basis with other people, walls, furniture and all other items in the office. Humidity, temperature, and draughts have little impact on radiation. It simply relies on a temperature difference between the individual's skin and the surrounding surfaces. If people work in a cool environment, they lose a lot of radiant heat throughout the day. This starts to be uncomfortable if they work alongside a cold window or a north-facing wall, even in what is considered to be an acceptable temperature. In this situation, an individual can be exposed to an asymmetric thermal environment, where the side of the body adjacent to a window is cool, but the other side is comfortable. Working alongside cold surfaces can account for about 40% to 60% of the total amount of heat lost in a day. Windows in modern offices tend to be larger than those in older buildings, so modern offices tend to have more difficulty: the larger windows present large cooling surfaces in the winter, and in summer, the same windows allow a great deal of heat to enter the room from the outside. If the mean temperature of an adjacent surface, such as a work surface, differs in temperature from the air temperature by about 2° or 3°, this will have an impact on how the individual perceives the temperature in the whole working environment. If an object, such as a wall, is about 4° colder than the air temperature, this will have a similar impact.

Temperature is the most often complained about disruptive factor in an office (Frontczak and Wargocki 2011). However, when considering working temperature, there is a vast difference between what would be considered acceptable in an area of the office where everyone sits down while they work, and areas where people are standing and physically active, such as in a mail room or archive room. Temperatures between 20°C and 26°C (68°F–78°F) are usually suggested as a suitable range for sedentary workers, whereas temperatures around 18°C to 19°C (65°F–68°F) are considered more suited to individuals engaged in more physically demanding work. What is viewed as acceptable by an individual will also be influenced by the clothing worn, the time of year, body size, gender, age, whether they have consumed food, and what type of food.

Exposure to draughts will influence an individual's perception of the temperature setting in their office. Difference in temperature values between the floor and around head level will cause discomfort. Differences in vertical temperatures can become more noticeable if the temperature is not distributed evenly by the heating/cooling or ventilation systems in use. Heat dissipation from equipment can also cause changes in vertical temperature, as can cold draughts along cold surfaces lower down in the workspace. A draught originating from behind the body will be viewed as less acceptable than a draught originating in front of the body. The ankles and neck are more susceptible to draughts than other parts of the body and the colder the draught, the less pleasant it is considered to be by the individual. The less movement involved in the individuals work, the more susceptible they are to experiencing draughts as being unpleasant. If people move around, they are less likely to notice draughts, and, if they do, they will not perceive them to be as problematic. The lower the ceilings in an office, the more likely the individuals working there will feel that they need to allow fresher air in to that environment, and they may find that it is warmer than one with higher ceilings. Draughts can originate from air-conditioning units or ventilation systems, especially if they kick-in violently to reduce the overall room temperature.

Humidity also has an impact on thermal comfort. If the humidity levels are low, the air will be dry and people may experience itching of the nose and eyes, and contact lens wearers may find that their lenses start to irritate their eyes. This is a common problem in offices owing to the amount of equipment in use that gives off heat and dries the air. The problems are worse in winter when heating is employed. A relative humidity of 40%–50%, combined with the moderate range of temperatures previously discussed, will probably be satisfactory for most sedentary office workers wearing light clothing. It is likely that the upper ranges of the 20°C–26°C moderate range will be too high for some individuals, who may become drowsy. Higher temperatures are also likely to result in increased heart rate, increased blood pressure, reduced activity of the digestive system, a significant increase in skin temperature, a large increase in blood flow through the skin, and more profuse sweating. In extreme situations, as the overriding goal is regulating body temperature, all other systems must be pushed into second place. As a consequence, the muscles will not have the same level of blood supply and will work less effectively. In addition, the digestive system has a reduced blood supply, which might result in nausea, as the stomach sends a signal that it does not want any intake of food.

It is generally believed that temperatures around 21.5°C to 22°C represent a satisfactory level for office workers. If the temperature is maintained within a range that would be considered 'normal', people will not notice it, because they will be comfortable. The more it deviates from the normal range, the more likely it is that people will notice that it is not a comfortable environment for work. In extreme conditions, some individuals may start to experience discomfort, or even pain, which prompts them to take action, such as removing some clothing or adding additional layers of clothing. If people do start to feel uncomfortable because the environment has moved outside a suitable range, they will work less efficiently. If they feel too cold, they will become fidgety and be unable to concentrate. They may experience drowsiness if the environment is too warm. The dissatisfaction caused by an elevated ambient temperature has been shown to reduce cognitive performance (Gaoua et al. 2012). Users are more likely to make mistakes because they are not as alert. Working in warm environments is exhausting and Syndicus et al. (2018) are of the view that it lowers our willingness to exhibit high levels of work performance and perseverance. This result would suggest that decrements in persistence might have an impact on decision-making by preventing a thorough evaluation of the decision outcomes, or choice alternatives, therefore leading to suboptimal decisions. It is likely that there would be similar outcomes in areas like problem-solving, innovation, and creativity, where a neglect of alternative ideas and a premature closure of the idea generation process would lead to suboptimal results.

Obviously, the clothes an individual wears will have an impact on their perception of how acceptable the environment is for the work they do. Once the individual wears clothes, they create a microclimate between the skin and the clothing. This microclimate should allow them to maintain a satisfactory heat level by maintaining skin temperature and permitting the required amount of sweating. The insulating qualities of clothing will be affected if it gets wet, as through sweating or from rain. Moving around, ventilation and penetrating wind will also affect the clothes' insulating qualities, which tend to 'pump out' warm air from between the skin and the clothing. It is generally recommended that clothes should allow pumping and venting at the neck, waist, wrists and ankles, so trapped air can be exchanged if the individual is getting hot. This may have implications in warm environments for males, for example, where they may be expected to wear ties, belts on their trousers, and have snug-fitting cuffs.

If measurements are being taken in the workplace to confirm whether the thermal environment is acceptable, readings should be taken in as many places as possible to ensure accuracy of the data collected. Readings should be taken alongside workers and should include measurements at ankle, chest, and head height. Measurements should also be taken at intervals throughout the day, because the climate may change for a number of reasons.

Nicol et al. (2012) believe that as energy is becoming more expensive and the climate more extreme, we require a new approach to meeting the challenge of designing comfortable buildings in which people can work. They have suggested an adaptive approach to thermal comfort, which starts from the observation that there are a variety of actions people can take to achieve thermal comfort, and that discomfort is caused by constraints imposed upon those actions by social, physical, or other factors.

9.5 Summary

- Noise is defined as unwanted sound.
- Noise is a subjective experience.
- Sound can be measured objectively.
- Sound does not have to be loud to be considered noise.
- Sound becomes noise when it breaks concentration, is distracting, is annoying, or is unacceptable to the individual.
- Persistent noise can result in anger and frustration.
- The nature and context of noise are more important than the intensity of the noise.
- Noises are more distracting in a quieter environment than in a noisier one.
- Noise is more annoying if it varies, if it is high-pitched or deep and rumbling, if the content is meaningful, and if it is under the control of another.
- Noise is less annoying if it is predictable and is viewed as being an unavoidable and necessary part of the work.
- Performance can deteriorate as a result of continued exposure to noise.
- Open-plan offices can lead to a deterioration in performance and dissatisfaction due to noise and lack of auditory privacy.
- Activity-based offices may offer an improvement over open-plan offices as they to provide a diversity of settings allowing people to move to a quiet setting.

- Reverberation has an impact on noise levels. Reverberation refers to the bouncing of sound waves off surfaces.
- The fewer the sound-absorbing surfaces, or objects in a room, the greater the reverberation.
- Carpeting and partitioning act as sound absorbers.
- Background noise is most distracting if it shares similar frequencies with the target sound. This is referred to as masking.
- Noise levels should be controlled to maintain performance at acceptable levels and reduce annoyance.
- The levels of light required to complete a screen-based task efficiently are different from those needed to complete a paper-based task efficiently.
- Luminance relates to light bounced back by a surface.
- Illuminance relates to the amount of light falling on a surface.
- High levels of illuminance cause glare.
- Glare can be direct or indirect. It occurs when there is an imbalance of surface luminance within the visual field.
- Disability glare makes completion of a task more difficult.
- Workers should sit between rows of lights rather than in line with them.
- Workers should sit parallel to windows rather than facing them or sitting with their backs to them.
- Lighting levels can influence workers' mood.
- 500 lux is suitable for screen-based work.
- 700 lux is suitable for paper-based tasks.
- Desk lighting offers greater control to the user over the local lighting environment.
- Cleaning lamps or their shields, replacing failed bulbs or tubes, changing room surfaces from dark to light, moving light sources together, and adding more may improve lighting conditions so they become acceptable.
- People prefer to work in rooms with windows rather than without.
- Workers should be able to see sky through a window.
- Tall windows are more effective than broad ones.
- People's bodies react to maintain their core temperature.
- Although a range of 20°C–26°C is considered acceptable, temperatures between 21.5°C and 22°C are most likely to be considered satisfactory.
- Draughts influence a worker's perception of the temperature.

- Differences in temperature between the head and ankles will cause discomfort.
- Draughts behind the body are less acceptable than draughts in front of the body.
- The ankles and neck are most susceptible to draughts.
- The less a worker moves, the more noticeable the draughts.
- Low humidity causes irritation of the mucous membranes and eyes.
- Wearing clothes creates a microclimate between the skin and the clothes.
- Measurements of the thermal environment should be taken at different levels and at different points during the day.

References

Banbury, S. P., and Berry, D. C. 2005. Office noise and employee concentration: Identifying causes of disruption and potential improvements. *Ergonomics* 48(1), 25–37.

de Croon, E. M., Sluiter, J. K., Kuijer, P. P., and Frings-Dresen, M. H. 2005. The effect of office concepts on worker health and performance: A systematic review of the literature. *Ergonomics* 48(2), 119–134.

Evans, G. W., and Johnsson, D. 2000. Stress and open-office noise. *Journal of Applied-Psychology* 85, 779–783.

Frontczak, M., and Wargocki, P. 2011. Literature survey on how different factors influence human comfort in indoor environments. *Building and Environment* 46, 922–937.

Gaoua, N., Grantham, J., Racinais, S., and El Massioui, F. 2012. Sensory displeasure reduces complex cognitive performance in the heat. *Journal of Environmental Psychology* 32(2), 158–163.

Helenius, R., Keskinen, E., Haapakangas, A., and Hongisto, V. 2007. Acoustic environment in Finnish offices e the summary of questionnaire studies. In: *International Congress on Acoustics*, September, Madrid, Spain. Sociedad Espanola de Acustica (SEA).

Jahncke, H., and Halin, N. 2012. Performance, fatigue and stress in open-plan offices: The effects of noise and restoration on hearing impaired and normal hearing individuals. *Noise Health* 14(60), 260–272.

Jahncke, H., Hongisto, V., and Virjonen, P. 2013. Cognitive performance during irrelevant speech: Effects of speech intelligibility and office-task characteristics. *Applied Acoustics* 74(3), 307–316.

Kaarlela-Tuomaala, A., Helenius, R., Keskinen, E., and Hongisto, V. 2009. Effects of acoustic environment on work in private office rooms and open-plan offices – longitudinal study during relocation. *Ergonomics* 52(11), 1423–1444.

Kim, J., and de Dear, R. 2013. Workspace satisfaction: the privacy-communication trade-off in openplan offices. *Journal of Environmental Psychology* 36, 18–26.

Küller, R., Ballal, S., Laike, T., Mikellides, B., and Tonello, G. 2006. The impact of light and color on psychological mood: A cross-cultural study of indoor work environments. *Ergonomics* 49(14), 1496–1507.

Laurance, J. 2006. Why artificial light could brighten up your day, The Independent; Health Medical Section.

Liebl, A., Haller, J., Jödicke, B., Baumgartner, H., Schlittmeier, S., and Hellbrück, J. 2012. Combined effects of acoustic and visual distraction on cognitive performanceand well-being. *Applied Ergonomics* 43, 424–434.

Nicol, F., Humphreys, M., and Roaf, S. 2012. *Adaptive Thermal Comfort: Principles and Practice*. Abingdon, UK: Routledge.

Passero, C. R. M., and Zannin, P. H. T. 2012. Acoustic evaluation and adjustment of an open-plan office through architectural design and noise control. *Applied Ergonomics* 43, 1066–1071.

Rea, M. S. 2000. *The IESNA Lighting Handbook*. 9th ed. New York: Illumination Engineering Society of North America.

Rolfö, L., and Eklund, J. 2015. "Examining office type preference." *Paper Presented at the Nordic Ergonomics Society 47th Annual Conference*, Lillehammer, Norway.

Rolfö, L., Eklund, J., and Jahncke, H. 2017. Perceptions of performance and satisfaction after relocation to an activity-based office. *Ergonomics* 61(5), 644–657.

Seddigh, A., Berntson, E., Danielson, C. B., and Westerlund, H. 2014. Concentration requirements modify the effect of office type on indicators of health and performance. *Journal of Environmental Psychology* 38, 167–174.

Sheedy, J. E., Smith, R., and Hayes, J. 2005. Visual effects of the luminance surrounding a computer display. *Ergonomics* 48(9), 1114–1128.

Syndicus, M., Wiesea, B. S., and van Treeck, C. 2018. Too hot to carry on? Disinclination to persist at a task in a warm office. *Environment Ergonomics* 61(4), 476–481.

Wohlers, C., and Hertel, G. 2016. Choosing where to work at work – towards a theoretical model of benefits and risks of activity-based flexible offices. *Ergonomics* 3, 1–20.

Woodson, W. 1981. *Human Factors Design Handbook*. New York: McGraw-Hill.

10

Manual Handling

10.1 Introduction

One of the biggest mistakes that can be made by people working in offices is to believe that because they are sedentary workers, they will not be involved in manual handling activities. Everyone in an office becomes involved in manual handling tasks in that environment at some time. For instance, boxes of copier paper are moved to the photocopier, archive boxes are filled and moved to a storeroom, or rooms are set up for training courses or seminars, resulting in desks and chairs being moved, possibly along with audiovisual equipment. Many office workers move files around the office, and many carry laptop computers to and from work. Large bottles of water will be placed on the drinks dispenser at intervals, and workstation furniture may be relocated within the office. The movement of all of these items is manual handling, and, more often than not, the individuals given the responsibility of moving them have not received manual handling training. It is likely that they have not received training because of the erroneous assumption that, as office workers, they will not be handling loads. Any individual moving objects within the office environment should be viewed as being at risk if they have not received the training to complete this task properly and safely. They will have little understanding that sometimes they will be moving quite heavy loads—possibly load weights heavier than some handled in industry. For instance, a laptop in its bag can weigh 9 kg, and a box of copier paper can weigh around 13 kg. Archive boxes are commonly filled with paper weighing in excess of 15 kg. A large bottle of water used in a dispenser can weigh around 18 kg. Even a bundle of files removed from a filing drawer can easily weigh 7 kg or more. Office handlers do not realise that when they carry bundles of documents around the office they are, in effect, moving a lump of tree. When viewed in these terms, it is easy to accept that they are involved in quite heavy manual handling at times.

Everyone in an office should be given manual handling training and it should be tailored to suit the type and amount of manual handling they might perform. Those working on the periphery of the main office departments should not be overlooked either. For example, the mailroom is a hotbed of manual handling, as sacks of mail and parcels are delivered to the site and sorted. Many people who have never ventured into the mailroom in their building will be surprised to see that bags of mail are typically thrown or swung around the work area. They will see mail being sorted, then pushed around the various departments in an assortment of trolleys, some of which may not be functioning properly, which makes moving them more demanding.

Individuals who work away from the main office, because they work remotely or because they travel between offices in different areas or travel to conferences, etc., also need to be given manual handling advice. Business drivers who drive for 20 hours or more per week are considered to be particularly susceptible to experiencing symptoms of a musculoskeletal disorder (Porter and Gyi 2002), and one of the reasons suggested for this susceptibility includes manual handling risks (Basri and Griffin 2012). Harris and Mayhog (2003) have previously suggested that priorities for reducing musculoskeletal disorders amongst business drivers should include the introduction of a robust manual handling policy for laptops and other equipment, in addition to risk assessment, and a management approach which involves both the employer and the employee.

Within the confines of the office building, there are other areas that require attention when focussing on manual handling. For instance, if the building has its own restaurant or café, a significant amount of manual handling will occur in the kitchen. As an example, a bag of potatoes can weigh 25 kg and will often be handled by female kitchen staff without assistance. Similarly, security staff probably sign for parcels in the reception area of a building and then are responsible for moving them to the appropriate department. IT personnel may be responsible for collecting and relocating IT equipment. Cleaners who move around the building before or after general working hours have to move mops, buckets, vacuums, cleaning materials, and so on, all of which can be heavy.

Having drawn up a list of who might move objects within an office building, it is easy to see that almost everyone could potentially move something at some time. For that reason, all workers need to be offered the right type of manual handling training programme.

Many who do recognise that they are involved in manual handling may believe, inaccurately, that only handling very heavy loads will harm them. On that basis, they might feel that moving a couple of reams of copier paper or a handful of files will have no impact on them, no matter how they carry them. They need to realise that what causes problems during manual handling is the way it is done, not necessarily the weight of any particular load. Repeated handling of even modest loads can cause back problems if performed over a lengthy period.

If a motivator is needed to prompt an organisation to take its manual handling issues in hand, thought should be given to the impact on individuals if they become injured when handling a load at work. On a personal level, the instant impact will be pain, discomfort, and/or restricted movement. After that, it is likely there will be a change in the individual's quality of life. They may find that everyday tasks, such as walking upstairs, getting out of a chair, or putting on socks, become difficult, painful, or even impossible. This is likely to be followed by a change in their financial circumstances as they recover from injury, and further still if they cannot return to employment. If they do return to some form of employment, their choices may be limited by their injury, and they may need to accept a job with a lower pay. They may also have to pay to be retrained in a completely different field. Other costs may include making alterations to the home to make it easier to move around, and they may have to pay for medication and treatments such as physiotherapy.

The cost to the organisation if someone is injured at work should also be a strong motivator for confronting manual handling issues in the workplace. The immediate impact of a manual handling accident is loss of input from the injured party and from any colleagues who stop work to assist them. If time has to be made up through overtime, this will be costly to the organisation. In addition, while the injured person is absent, their colleagues will probably have to cover for them, which increases their own workload on a day-to-day basis. Any accident investigation will involve time, as will any report that has to be prepared. There are likely to be costs associated with the administrative side of the injured person's absence, such as replacing the individual and retraining a replacement. The company may also have to pay medical expenses, such as specialist's fees or physiotherapy sessions. There may be issues relating to any future personal injury claim by the individual, and this is likely to have an impact on insurance premiums.

10.2 Manual Handling Injuries

If people were asked to list the areas of the body likely to be hurt in a manual handling accident, most would probably suggest the back first of all. They are correct that, in most cases, the back is the most likely body part to be injured, but it is not the only body part they can hurt. Many people develop upper limb disorders (ULDs), especially if they handle loads repeatedly at work. ULDs are dealt with in Chapter 11. Other handlers may drop loads on their fingers and feet and these, too, are classed as manual handling injuries.

The structure of the back and some of the reasons why it might become injured are dealt with in Chapter 1. In brief, it is possible for damage to the discs to occur over time if the individual leans forward at the waist while lifting a load. This causes bulging of the disc, and, over time, this bulge can increase in size until it presses on the nerves, causing discomfort. Occasionally, an individual will experience a prolapse. In addition, when a person leans forward, the muscles also have to combat the effects of gravity; controlling the weight of the upper body when in this posture can overload and injure the back muscles.

Ligaments run almost the full length of the vertebral column. They are passive links between bones, in contrast to muscles which provide active links between bones. The ligaments in conjunction with the muscles and joint capsules bring about movement. The ligaments can become stressed by repeated bending at the waist, which stretches them, and by twisting at the waist. Twisting at the waist is usually unnecessary. People simply have to be aware that they should move their feet when moving an object, rather than swinging it from one surface to another without turning their feet in either direction. Moving the feet will immediately eliminate twisting at the waist.

Hernias are often associated with manual handling, and it is usually assumed that these must be the result of heavy manual handling. This is part, but certainly not all, of the picture. Hernias occur most commonly in the abdomen. The abdominal wall is comprised of tough muscle and tendon that runs from the ribs to the groin, principally holding the intestines in place. The intestines exert significant outward pressure even if the individual is not engaged in manual handling. Sometimes a person will have a weakness in the abdominal wall. As they raise a load off a surface, there is an increase in intra-abdominal pressure, which, in effect, pushes the intestines forward against the abdominal wall. Sometimes a piece of intestine pushes against and through the weak point, resulting in a bulge—a hernia. If investigations are carried out after the development of a hernia, it is likely to be established that the individual had a pre-existing weakness in the abdominal wall, they were involved in heavy manual handling that would have increased the internal pressure, and they combined this with poor posture, such as leaning forward at the waist while lifting the load. Leaning forward would have increased the pressure on the abdominal wall even more.

10.3 Reducing the Risk

The most effective way of reducing injuries is to reduce the risk of the manual handling task. The hazardous elements of the task can be identified during a specific manual handling assessment, which will be discussed in Chapter 13 on risk assessment.

In the first instance, thought has to be given to whether the load needs to be handled at all. For instance, instead of having employees carry boxes of copier paper to the photocopier, the supplier could deposit it in a storage unit next to the copier. This would mean that employees only need to pick up one or two reams of paper as required. Instead of archive boxes being carried to storerooms, the employees could use mechanical assistance, such as trolleys, to transport the boxes.

10.3.1 Mechanical Assistance

Although using a trolley lessens the burden of handling a load, it should not be forgotten that using it retains an element of manual handling as the load is pushed or pulled. The condition of the trolley affects how easy the task is; therefore, it is essential that all trolley or other mechanical equipment used for transporting loads should be well maintained. The wheels on a trolley often become worn or damaged, and if they interfere with the free movement of the load, the individual pushing it will have to overcome the resistance by increasing the force applied.

Providing any type of mechanical handling device will not necessarily make the task easier. The equipment, such as a trolley, has to be selected with both the task and the environment in mind. The trolley itself should be designed so that it can manage the load easily, and this will require considering the style and number of tiers incorporated into the design. Workers will have to learn how to stack a trolley properly so that the load remains stable during movement.

The wheels on a trolley should suit not only the load weight being transported, but also the surfaces across which the trolley will be moved. Some wheels are designed to traverse vinyl, others concrete, others carpet. All wheels will not work equally well on all surfaces. Nylon or cast-iron wheels work most effectively on hard, smooth surfaces such as concrete, but they are not good at traversing gaps in the floor or bumps. Using rubber materials in the wheels makes them easier to move over a bumpy surface or outdoors, as the rubber absorbs some of the shock. Pneumatic tyres are effective when moving over uneven surfaces, but they need to be monitored regularly to ensure that they have the correct pressure.

The size of the wheels and the number present also have a bearing on how easy a trolley is to push or pull. Larger wheels are better for use in areas where there are gaps in the flooring or the floor surface is not particularly smooth, as they have lower rolling resistance. Trolleys with smaller wheels are acceptable for use with lighter loads and for moving only a short distance. Narrow wheels and types with a rounded profile move more easily on hard surfaces. A wider type is necessary if there are gaps in the flooring where a narrow wheel might get stuck, or where the area is carpeted. If people have to use the trolley in a variety of areas with either carpeted or concrete flooring, they should use a trolley with a soft tyre and a solid central rim.

Two-wheeled trolleys, such as found on a sack truck, can be used in a variety of situations, ranging from the movement of several boxes of copier paper to moving a filing cabinet. However, as the trolley is supported by the handler once it is tipped back, these are not suitable for movement across long distances. Platform trolleys might be a better alternative and are also more suited to the movement of larger, bulkier items. Handlers need to be aware that they should avoid bending if placing smaller items on the low-level platform. This problem can be overcome to an extent by using platform trolleys with an upper shelf for smaller items. Having drop-sided shelves, where a retaining bar or gate can be lowered, will make an item more accessible when being removed from the shelf.

Whether the wheels have been designed to rotate on the front or rear of the trolley, and how easy this makes the trolley to steer, has to be considered. Four swivelling wheels are most suited to confined or congested areas because they allow the trolley to be manoeuvred more easily. However, they should only be used over short distances, because they require greater effort to control and steer. Care should be taken if they are used on sloping surfaces, as the trolley may start to move sideways, which the handler will have to resist. As a general rule, when moving up or down a slope, the handler should be uphill of the trolley so that any loss of control does not result in the trolley colliding with them. A trolley with four swivelling wheels should be fitted with a handle at each end. A combination of two fixed and two swivelling wheels is more appropriate for longer distances and for slopes. Slopes with a gradient of more than 2% will result in a significant increase in effort on the part of the handler to control the trolley during ascent and descent. Trolleys frequently used on slopes should be fitted with brakes, which can be either hand- or foot-operated. Whatever type is used, it should not require excessive force to operate, and foot brakes should not be positioned so that they are likely to come in contact with the handler's legs or feet as they walk. If the trolley is very long, two central fixed wheels and two swivelling wheels at each end will make it easier to steer around corners, but will make it difficult to park neatly against a wall when not in use.

Where users place their hands when pushing the load will have an impact on the effort involved. Being able to grip a well-designed handle between waist and shoulder height is desirable, as this allows the handler to stand upright and not have to work with her hands outside a comfortable zone. Using vertical handles, rather than horizontal handles, specifically on narrow trolleys, allows individuals to place their hands at the most convenient height for them. These should be positioned about 450 mm apart. Horizontal handles do make manoeuvring a trolley in confined spaces an easier task. Cylindrical handles are more appropriate and should have a diameter around 25–40 mm. Covering the handle in insulating material, such as rubber, will reduce the possibility of heat exchange and the likelihood of the hands slipping during use. There should be sufficient clearance when handlers push a trolley through a doorway so that their hands do not become trapped or crushed. Ensuring that trolleys are about 80 mm narrower than the smallest door aperture used should provide this clearance. If doorways cannot be used easily, the handler will feel compelled to abandon the trolley and carry the items being transported, which defeats the purpose of providing the trolley. If handles are fitted to the trolley, they should be positioned inward from the outer edges of the trolley to ensure that hands do not get injured when going through narrow passageways.

Observations of handlers should identify whether they use only one hand to control the trolley, such as pulling it behind them. The amount of force that can be generated by one hand is only about 50% to 60% of that which can be generated with two hands. Pulling one-handed also causes the handler to twist to one side and the workload is focussed on one side of the body. Pushing or pulling with two hands spreads the workload out more evenly across the body and permits the handler to maintain an upright, symmetrical posture. If the load height permits it, handlers should always push the trolley in front of them. A trolley height of 1400 mm should ensure that even a '5th percentile' female handler will be able to see over the trolley, given an unshod eye height of 1405 mm (McKeown 2011). This height also relates to stacking levels on the trolley. This height should increase stability of the trolley when it is being moved up a slope, since very tall trolleys can tip up when pulled up slopes. If the trolley is constructed so that it has high, solid sides, consideration should be given to replacing the upper level with mesh or cross-frames at appropriate points to enable the handler to see ahead.

It should not be assumed that because a trolley is being used, the handler will not have to employ much force to move the load. The only satisfactory way to confirm that excessive force does not need to be used to move the load is to measure the pushing or pulling forces. When taking such measurements, two different types of readings need to be recorded. First, the force required to start the load moving needs to be recorded. Subsequently, the force required to keep the load in motion should be recorded. Typically, more effort is required to start the load moving than to keep it moving. Sample measurements should be taken in a number of areas that are representative of where the load would be moved. Therefore, this might require taking measurements in carpeted areas, where carpets adjoin vinyl floors, where handlers might turn a corner, where they might pull a trolley into an elevator, where they might pull the load up a slope, and so on.

Should the effort involved in moving the load be excessive when compared with applicable standards and guidelines, the employer will either have to reduce the total load weight being transported or try an alternative trolley with different wheels more suited to that environment, or both.

The footwear chosen by people who work in offices may increase the difficulty of pushing and pulling a trolley owing to the style and materials of the shoes. The floor surface also plays a part in the difficulty of the task. Where possible, handlers should be able to push or pull trolleys along even, dry floor surfaces. The use of automatic opening doors will make movement between different work areas less demanding.

10.3.2 Work Demands

If someone has no choice but to handle a load without any form of mechanical assistance, an evaluation needs to be made of how the task is completed to identify whether the risk of the operation can be reduced. The posture adopted when moving the load should be considered. It is always recommended that a load be kept close to the body; therefore, if the handler holds or manipulates the load at a distance from the body, the reason for this should be explored. For example, it may be that when handlers load boxes of copier paper onto a shelf, they have to reach across archive boxes stored on the floor directly in front of the shelves. This will cause them to work with their arms extended and possibly leaning forward. Holding a load at a distance from the body is five times more stressful than holding it close to the body. Relocating the archive boxes will allow the handlers to get closer to the shelves and will reduce the extent of reaching or leaning forward, which will reduce the risk of injury. Handlers should be made aware of the need to get as close to a load as possible before attempting to pick it up. For instance, rather than lifting a monitor at a distance on the desk surface, they should slide it forward across the desk surface before picking it up. The correct techniques to employ when handling loads are explored more fully later in this chapter.

If handlers twist at the waist as they transfer items from one surface to another, such as from a trolley to a shelf, this will increase the risk of backache or back injury, and this is usually unnecessary, as the handlers are likely to be completely mobile. They should move their feet to face the direction in which the load is moving. Even individuals sitting on seats have a tendency to twist at the waist when they want to remove an object, such as a reference folder, from an adjacent shelf, rather than rotating the swivel chair so that they face the shelf and avoid twisting.

People who are unaware that they are involved in manual handling, which is typical of office workers, frequently bend forward at the waist as they try to raise something from a low surface, such as the floor or the lowest drawer in a filing cabinet. This is particularly stressful for the back, as the upper body becomes a 'load' that also has to be controlled during the movement of the intended object. Apart from training handlers in the correct technique, consideration should be given to storing or presenting loads at heights that reduce the need (or at least the temptation) for handlers to lean forward. Presenting items between knuckle and elbow height is considered ideal, particularly if the loads are heavy or awkward. Higher and lower shelving can still be utilised, but these shelves are better suited to the storage of lighter or infrequently used items. Figure 10.1 gives a clear indication of the lifting ability of males and females. This can be used as a guide when determining the positioning of loads on shelves or other surfaces. If shelves are designated as heavy load or light load shelves, they should be marked or labelled in some way so that the handler is aware of what can and cannot be placed on them, such as when unloading a trolley. Marking the loads themselves with an indication of weight will also be helpful.

Removing objects from high surfaces, or placing them there, is demanding for both the arms and back. The arms are working at a mechanical disadvantage, making it more difficult for them to control the load, which could result

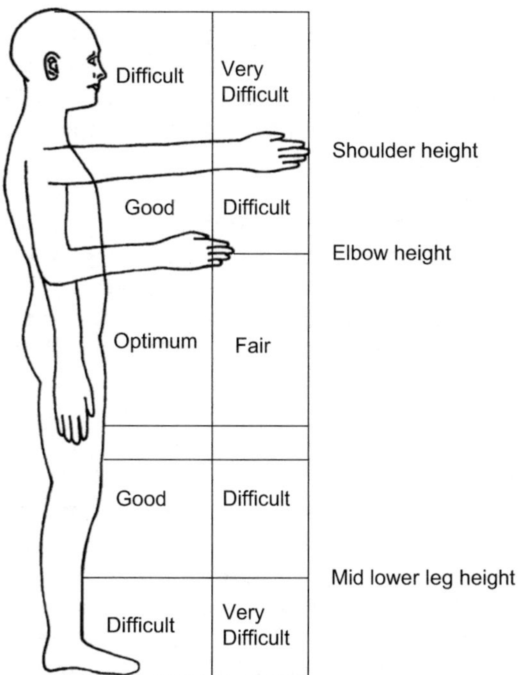

FIGURE 10.1
Lifting ability relative to height and distance from the body.

in arm or shoulder injuries. Moving the load through a large vertical distance, such as from the floor to the top of a filing cabinet, should be viewed as a demanding task, because it requires both a change in posture and a change in grip as the load is moved. Ideally, heavy loads should not be stored on or near the floor, and they should not be stored at high levels, such as above shoulder height. If people do have to move loads from a high point to a low point, or vice versa, they should use a halfway surface around waist height to rest the load so that they can change their grip and posture appropriately before finishing the movement. Although it might be argued that this results in two separate lifting actions, rather than one, it is always better to do two safe lifts rather than one unsafe lift.

If the load is being positioned precisely, such as when placing a file back in a filing drawer, the length of time the handler is in contact with the load is extended, and this will increase the fatiguing effects of the handling operation. If the load is being removed from its point of storage and is released abruptly, such as a box suddenly being pulled free from a shelf, the handler may be taken by surprise. Being unprepared for the movement of the load may result in the handler becoming injured.

People who are involved in continuous handling of small loads should be viewed as being at risk, particularly if they are rushing. It is important that handlers understand that it is not just large, heavy loads that cause difficulties. Regular breaks from handling allow people to recover from the effects of continuous handling. Allowing handlers to choose when they need to stop for a short break is more effective at controlling the onset of fatigue than setting specific intervals for breaks.

If people are sitting down when they handle a load, such as handling archive boxes as they sort through the documents inside, their ability to lift safely is probably reduced by about 80% when compared with handling in their optimum standing position. This is because once they sit down, they are fixed in place and cannot move toward the load. As a consequence, they become completely reliant on their arms and upper body to do the work. If they are sitting on a standard office chair on casters, they should be aware that the chair might move backward if they reach forward causing them to fall. Seated handlers should avoid trying to lift something from the floor while remaining on the chair as this will cause them to stoop. For that reason, files should not be stored on the floor area around a worker. If there is insufficient space on the desk surface, a temporary surface, such as a trolley, could be used alongside the seated worker.

The greater the distance an individual carries a load, the greater the demands of the task. If someone carries an object for more than 10 meters, they are likely to become increasingly fatigued, making them more vulnerable to injury. Providing a trolley would help in this situation. If so, the next thing to consider is whether there is room to move the trolley freely and whether there are steps and slopes that would make the task more difficult.

Occasionally, two or more people may be required to move something large or heavy. It should not be assumed that two people can lift twice as much as one person, or that three people can lift three times as much as a single individual. Once people start working together to handle loads, there are additional burdens that increase the risk of the operation. For instance, one person may need to walk backwards when handling the object, increasing the risk of tripping. Once the second handler takes up position on the opposite side of the load from the first, they may impede each other's view. Having two or three people standing around a load as they transport it increases the space requirement in terms of clearance through doorways and along corridors. Sharing the handholds can be more difficult as more people assist with the lift, forcing some of them to grip the load in more awkward positions. If the load is very small, the handlers might find there is insufficient space for all of them to grip it without overlapping with each other. An asymmetric load may result in one of the handlers having to bear a greater proportion of the weight than another in the team. In this situation, the strongest team member should take the heaviest end. Handlers may not communicate properly with each other with regard to raising and lowering the load resulting in an uncoordinated lift. This emphasises the need for appropriate training in team handling, as well as general manual handling skills. Generally, if a load requires more than three people to move it, an alternative non-manual method should be employed. If two people work as a team, it should be assumed that they will probably have about two-thirds of their combined individual capability, and three people working as a team will have about half of their combined individual capability. People in the team should be around the same height and about the same level of ability, if possible.

10.3.3 Object Characteristics

People usually focus on the weight of a load, initially, in the belief that it is likely to be the major source of any difficulty they will have transporting it. They often assume that the heavy loads are likely to be the most problematic. However, that does not always prove to be the case. Handlers are often cautious when faced with a large heavy load and take extra care when moving it. They tend to treat it with respect, and, as a consequence, they probably reduce the risk of injury. When faced with a smaller, lighter load, some handlers will take a more casual approach, possibly neglecting to use their learned manual handling skills. This attitude will put them at risk. It is common for people to think that a small object is so light that they can actually carry three, four, or more of those objects simultaneously. As a consequence, although the individual load weight might not be excessive, the total load weight will probably move into the excessive category when given a risk rating. For example, some people choose to carry one box of copier paper in each hand, holding them by the nylon straps. Instead of moving a load weighing around 13 kg, the individual is actually moving a 26 kg load which is significantly more demanding for the body.

Other features, apart from the weight of the load, may increase the risk of injury—for instance, its shape. If an object is large or bulky, such as a notice board, it can be difficult to grip. This is particularly so if it does not have any specific handholds, as is true of a television. Bulky loads can also make it more difficult for handlers to see where they are going or to see tripping hazards on the floor ahead. Using trolleys for such large objects might be a more suitable alternative. Floppy loads, such as large potted plants, are also more difficult to control.

If the load does not have a symmetrical distribution of weight, handlers may increase their risk of injury if they do not pick it up correctly. The heaviest side of the load should be held closest to the body. This may necessitate turning the object around or walking around to another side and approaching it differently. Many loads in an office, such as chairs and televisions, have an asymmetrical weight distribution. Appropriate training should advise handlers about this feature of loads, and they should be told to test the weight distribution of unfamiliar loads by rocking them from side to side before picking them up.

Some loads are more difficult and therefore more demanding to pick up because their outer surfaces are smooth or slippery, or they may be so large that they are difficult to grip. Loads like these should either be placed on a trolley for transport or placed within another container fitted with handholds to make them easier to move. If two people are working together to move such a load, they could consider using a sling-type arrangement, if a trolley is not available. For instance, if two female canteen workers were trying to move a 25 kg bag of potatoes, which will not have handholds, they could slide it onto a large refuse sack and pick it up using the corners of the refuse sack, which would act as a sling to support the load being transported; or, of course, they could slide it across the floor.

If loads are provided with handles, such as the holes cut into the sides of an archive box, it should not be assumed that the handles will actually serve a useful purpose. Some loads are better picked up without using the handles or straps. For instance, people often carry a box of copier paper using the nylon strap. Supporting the weight of a 13 kg load on the narrow strip of the palm covered by the nylon strap cannot lead to the use of a strong grip, which will be fatiguing, nor can it be healthy on a more local level for the palm of the hand. In addition, using one hand to support the weight of the load to the side of the body makes the handler lean or twist to one side. Handlers would be better to ignore the strap and lift the box by placing their hands underneath it. This would allow them to share the load between both upper limbs, they would maintain an upright, forward-facing posture, the load would remain close to the body when being carried, and they would be able to grip the box more firmly.

The manufacturer of the container in which a load is stored may not have thought through where the best place is for handholds. Manufacturers often work on a 'centralised' view of where handholds should be located. For instance, they usually place handholds in a central point on the sides of a box. This is not helpful if the box contains something like a microwave, with a distinct asymmetrical distribution of weight. It would be better to move these handholds in the direction of the heaviest side of the load, which would not only suggest to handlers where they should stand when lifting the load, but would also result in the heaviest side of the load being kept close to the handler's body.

Placing handholds on the top of a load can be quite useful if the load is small, as this will discourage users from bending forward at the waist to raise it off the floor. For instance, a laptop bag usually has a handle on the top so it does not have to be picked up by placing the hands beneath it at floor level. This type of handhold positioning might be more problematic if the load is big or long and will not clear the floor when being carried without the handler's arm being raised, which will be fatiguing for the limb. An example is a portfolio bag used by an artist or architect to carry large drawings. These often have handles at the top, but the bags are often so large that they drag on the ground if it is not raised high or carried differently, such as by using a handle fitted on the side so that the bag is carried almost tucked under the armpit.

Loads that are likely to shift when being transported increase the risk of a handling task. For instance, moving a large archive box filled with small objects, such as staplers and hole punches, may result in the objects shifting as the box is raised off the floor. This makes the load unpredictable and more difficult for the handler to control. Any load being transported within a container should be secured, where possible, to prevent it from shifting. For example, pushing a piece of bubble wrap or shredded office paper down the sides of a small object inside a large box will make it more stable. An item that is a little floppy, such as paper for flip charts, is difficult for the handler to control when moving it. Placing it in a rigid container or on a trolley will make it more controllable.

10.3.4 Environmental Conditions

The space in which people work when handling loads has an impact on how demanding the task is for them. Walking up steps or a ramp when moving between work areas is more demanding. If possible, all handling should be carried out on one level, and anything other than a very light object should be transported using the elevator rather than being carried up and down stairs. If handlers have to traverse two separate working levels using a ramp it should have a low gradient. The same principle applies to handling between different levels of shelves, such as in a storeroom or reference library. The level to which a load is transferred should be similar to that from which it has been removed to minimise the range of movement and thus reduce the risk of injury. On that basis, movements, such as transferring a box of stationary from the lowest level on a trolley to the highest shelf in the stationery cupboard, should be avoided through proper planning.

Narrow walkways, such as between rows of desks, or corridors influence how loads are carried or otherwise moved. Lack of overhead clearance, such as occurs when pulling something out from the lowest shelf in a storage area, affects the posture adopted and the subsequent level of risk. Having to negotiate doorways can make moving a load more challenging if the handler needs both hands to support it. Planning ahead—asking someone to hold the door open or placing a wedge under the door—is likely to happen only if the handler is provided with appropriate training to increase their awareness and, thus, their forward planning.

If handlers work in a warm environment, such as a kitchen or even a hot office, they may become fatigued rapidly, and may develop sweaty hands, which can result in them losing their grip on a load. If they have to work in cold environments, such as caretakers who have to clear up outside the main building, they may experience a loss in dexterity. If they are provided with clothing suitable for working in colder temperatures, such as bulky jackets and thick gloves, this may hinder their movement and interfere with their grip.

Working in well-lit conditions will ensure that handlers do not trip while moving between areas; they will not have to lean forward in order to see the load more clearly, and they will not have to spend so long putting the load on a surface if they can see exactly where they are positioning it.

10.3.5 The Person

Manual handling is probably one of the few subjects where legitimate distinctions can be made between people on an individual basis, particularly in terms of age, gender, and physical suitability. It is generally accepted that women do not have the same lifting strength as men. However, there will be some overlap, and there will be some women who can easily handle heavier loads than their male counterparts. Despite these possibilities, in order to offer protection to the whole workforce, it should be assumed that women have about two-thirds the strength capacity of men, and this needs to be taken into account when assigning manual handling tasks to workers. A woman's ability to lift and handle loads safely should also be viewed as diminishing during pregnancy. Although this does not necessarily suggest that women are incapable of handling loads safely when they are pregnant, expectant mothers are more susceptible to injury, owing to the combined effects of hormonal changes and their changing shape and posture, all of which affect their ability to lift safely.

In addition to the sensitive nature of dealing with gender when it comes to manual handling, the age of the handler also needs to be considered sensitively. Many countries are facing an increase in the age of their workers, because they are either remaining in full-time employment for longer or finding part-time work once they 'retire' from their full-time job. Research suggests (Finch and Robinson 2003) that the number of retirees re-entering the workforce is increasing.

Generally speaking, as humans grow older, age-related changes in their bodies can alter functional abilities and, as a result, this can increase the risk of suffering a musculoskeletal injury. Muscle mass decreases with age, with more dramatic decreases after 60 years of age. This loss in muscle mass has significant implications for muscle strength and endurance (Deschenes 2004; Mitchell et al. 2012). In addition, ageing has an impact on joint condition, including the quantity and quality of tissues, such as cartilage and ligaments (Loeser 2013; McCarthy and Hannafin 2014; Schleifenbaum et al. 2016). The reality is that age-related changes can influence the degree of joint pain experienced, as well as have an effect on flexibility and mobility. Businesses need to ask themselves if they have an up-to-date knowledge of the factors that are likely to impact on the health of their workforce, particularly as the make-up of that workforce is changing and will continue to do so for some time. Employers may need to consider if there are age-related differences that need to be accounted for when completing risk assessments. Chen et al. (2017) would question whether they also need to develop fluid or flexible risk reduction interventions that can be adapted to enhance work ability across their lifespan in the business.

Despite the difficulties associated with an ageing working population, it would be a mistake for an organisation to employ only young men in their late teens and early twenties to move loads around the business. Young men particularly will not have fully physically matured at this age and could be compromised if presented with a heavy or awkward load. Although, superficially it would appear that an older worker might also be more susceptible to injury, they bring with them certain qualities that offer some protection. For instance, older workers tend to have a more mature attitude to their work and as a consequence they may not feel the need to prove to anyone that they can handle heavy loads unaided. They will not be embarrassed to ask for help. In addition, older workers have a greater breadth of experience, which is likely to help them take a safe approach to handling objects. However, it has to be accepted that the older the workers, the faster they are likely to become fatigued and this will have to be accounted for when assigning work to them.

The clothing worn by office workers can potentially put them at greater risk of injury than their counterparts in industry. People working in industry tend to wear overalls or other loose workwear, and they have no concern about getting dirty. They are likely to wear safety shoes with steel toes and gloves that suit their work. People working in offices usually have to adhere to a dress code, which results in them wearing clothes that do not necessarily allow them to move easily, and they may be concerned about getting them dirty. Handlers are always advised to carry loads close to the body, but some office workers hold loads at a distance to avoid soiling their shirts. This concern over clothing immediately results in a five-fold increase in stress on the lower back. Female handlers who wear tight-fitting skirts and high heels may find it difficult to adopt an appropriate posture when handling loads.

Some individuals may suggest that they should wear a back support belt when handling loads in the office. It may be that they use one when they are working out in the gym. These need to be considered carefully because it is not believed that such belts offer any particular protection during load handling. In fact, they might increase the risk of injury because they generate a false sense of security, making the handler overconfident. If back belts are used, they should be worn properly. This means getting the right size to start with and ensuring that it is worn in the appropriate position on the body. To that end, the instruction leaflet that accompanies the belt should be read carefully before the belt is fitted. In addition, handlers need to be aware of what the belt is actually doing for them. It takes over some of the workload of the muscles in the area where it is fitted. If the body is being offered artificial support, the muscles may not offer the same level of assistance and support that they would normally as this would be duplication of effort. As a consequence, the muscles tend to be in a more relaxed state while the belt is being worn. The mistake that most handlers make is to take the belt off abruptly at the end of their handling session, which is 'shocking' for the muscles, which up to that point have been less active. To avoid the abrupt transfer of the workload onto previously relaxed muscles, the handler should gradually undo the back belt, one hole at a time, leading up to the point when they want to remove it. This allows a gradual resumption of effort on the part of the muscles as the belt is loosened.

It is important that the issue of self-selection is acknowledged when expecting individuals to perform manual handling tasks. Office workers have self-selected to work in what might be perceived to be a less physically demanding environment and have self-selected to avoid working in a more demanding industrial type environment where they might be readily expected to lift objects. On that basis, they may not be mentally prepared for the possibility that they have to handle loads and may not have the right kind of mind-set to approach it safely.

The key point with regard to handlers is that they are more likely to handle a load safely if they have been given appropriate manual handling training and awareness. No one should ever assume that someone knows how to handle an object correctly, unless it is certain that the individual has received suitable training and has developed the right kind of practical skills to do so.

10.3.6 Manual Handling Related to Vehicle Use

Employees who use their cars or other vehicles for work purposes will need some specific, tailored advice and guidance so that they can handle loads safely. More often than not, they will be working out of sight of their employer, and possibly on their own. For that reason, they need to have a very clear understanding of the importance of using safe handling techniques.

The following advice is based on many years of working with drivers in a bid to reduce the amount of musculoskeletal discomfort they experience. Training guidance on manual handling to and from vehicles is very thin on the ground and the list below has grown from the need to offer practical, usable guidelines to drivers who are usually left to work it out for themselves. It should be kept in mind that the guidance is not what would be considered ideal handling techniques. They are simply aimed at making a difficult situation less difficult.

A vehicle presents a handler with very tight and inflexible confines within which they often have to handle heavy and/or awkward loads. To reduce the risk in that particular environment, they should adopt the following principles:

1. The car should be parked as close to the drop-off/pick up point as possible. If parking is at a distance, they should phone ahead to get permission to pull in front of a building and close to the main doors so they can unload the items to be handled. This will reduce the carrying distance and the demands of the task, and this will, ultimately, reduce the risk of injury.

2. A brief walk after a long drive to loosen up a little before lifting heavy loads will be helpful.
3. If a load is being lifted by a seated driver from the front passenger area, such as the footwell, they should not reach across from their own seat to grasp the object. This would result in leaning, bending, twisting, and handling with the arms extended, and this increases the risk of injury. The driver should leave the vehicle and walk round to the passenger door to retrieve the object.
4. If items, such as laptops, are being taken out of the rear passenger area, the seated handler should not lean or reach through the gap between the front seats and drag the objects towards them. This will result in handling whilst seated, which reduces an individual's lifting capacity, and they will be working in a twisted posture, which will increase the risk. They need to get out of the vehicle and open the passenger door nearest the object they want.
5. One of the biggest elements of risk is removing objects from the rear passenger area when the handler is standing with both feet outside the vehicle. Reaching inside the vehicle through the low-level door opening often causes them to alter their posture to avoid coming in contact with the roof. More often than not, they will reach into the vehicle interior with their upper body leaning forward and unsupported. Objects should be slid towards the handler to get them close before they are lifted, if possible. If the handler is sorting through things, like a box of documents, or they are trying to reposition an object so that they can get a better grip on it, they could try resting one knee on the edge of the seat. This achieves several things: it lowers the height of the handler so their head may be lower than the roof line, it gets them closer to the object they are trying to pick up, they can remain in a more upright position, and it gives them support while they are getting organised for the lift.

   NOTE: This tactic could be employed by anyone trying to fasten a child into a car seat, as it will make the process less stressful, for the back at least. Although this may appear a simplistic piece of advice, people with children are likely to place their children in the car on a daily basis, maybe even multiple times per day, and the cumulative effect of adopting a poor posture on the back cannot be minimised. The situation is compounded if the child is a reluctant participant and fights the process of being placed in the seat, which extends the length of time the carer is standing in an awkward posture.

6. If the object in the rear passenger area is located further towards the centre of the car, the handler could try stepping into the footwell with one foot. This will get their upper body and arms closer to the

load, they will have a wider supporting base, which will make them more stable, and having the leading foot ahead will reduce the extent to which the upper body is in front of the legs, which reduces the gravitational effect. It also offers better support and stability if moving the load is delayed for any reason, such as trying to position the load for a better grip. If possible, the handler should reverse out of the car, sliding or dragging the load with them. They should, preferably, lift the load once both feet are flat on the ground outside the vehicle and the load is close to them.

7. If they are lifting items out of the boot, they need to stand as close to the boot/trunk as possible. They should stand with their feet apart for stability. They need to avoid keeping a distance from the rear bumper of the vehicle as a result of worrying about getting their clothes dirty. Handling loads at a distance from the body increases risk. Therefore, they either need to keep their car clean so they are happy to lean against it, which is unlikely to be easy given the distance they travel and the environmental conditions in which they will drive, or they need to keep something like an old sheet in the boot which they can throw over the bumper.
8. If the individual removes items that are heavy or awkward from the boot/trunk on a regular basis, the company should consider providing an estate car or station wagon, as these will not have boot/trunk sills to work over or around.
9. Anything impeding their view or hampering their free movement or access to the load in the boot/trunk such as a parcel shelf, should be moved.
10. If only their target item is in the boot/trunk, it is unlikely to be close to their standing position as it will have moved against the rear passenger seats on braking. It should be slid across the floor of the boot/trunk so that it is close to the handler before any attempt is made to pick it up.
11. If there are several items in the boot, they need to be carefully moved out of the way, preferably by sliding them to the side to avoid additional lifting.
12. If the boot/trunk has a sill or lip over which the load is lifted, it may be beneficial to lift the load from the floor of the boot and onto the sill first before repositioning the feet and changing the grip and lifting it

again. Having something over the top of the boot latch, like the old sheet, will stop the load snagging as it is temporarily put down.

13. If several items are being removed from the boot/trunk, or a single item is heavy or bulky, the handler should have access to a mechanical aid, like a sack truck or trolley. It may be beneficial for the handler to keep a lightweight, fold-up sack truck or trolley in the boot so that it always travels with them. If the handler is simply moving some documents and their laptop, a wheelie bag about cabin bag size would be advisable—unless the handler is moving over gravel or going up and down numerous stairs. The aim should be to limit or eliminate the need to manually carry or support the load, particularly over long distances.

    NOTE: Rucksacks for moving laptops are another good alternative, but, to be effective as a risk reduction measure, they should be worn over both shoulders, and any chest and waist straps should be buckled to stop the load moving around. The shoulder straps should be broad and padded to spread the load over a wide area. The rucksack needs to fit the length of the back properly. The material and frame of the rucksack should not make it heavy before it is filled. Heavier items should be placed towards the bottom of the rucksack. Messenger bags worn over one shoulder should be avoided, unless the items being carried are very light. The shoulder strap should be wide and padded and, preferably, the bag should be worn diagonally across the body.

14. If mechanical aids are not available, the handler should be decisive and make several trips back and forth between the car and the building when unloading.
15. Obviously, the reverse procedure is carried out when loading the car.
16. Working out of larger vehicles, such as vans, provides better opportunities for improved postures, as handlers may have increased head clearance and may be able to stand upright inside the vehicle. However, larger vehicles usually means larger loads, so greater care is needed. Sack trucks or trolleys should be provided for these drivers; not overlooking the fact that lifting mechanical aids in and out of a vehicle to use at different sites is also a manual handling task.

10.4 Training

The training offered to individuals involved in handling loads needs to provide an understanding of why it is important to learn about appropriate handling techniques and how those techniques should be applied. They should be given an appreciation of how parts of the body, such as the back or arms, can become injured during manual handling so they can appreciate the connection between poor handling skills and injury. Some trainees may comment on a course that they have been handling loads for many years without any training and have never encountered any problems, the inference being that they do not need any training at that time. What they need to understand is that many manual handling injuries are cumulative and build up over a long period of abuse. Therefore, although they may feel perfectly healthy at that moment, they cannot be sure they will avoid future problems.

If the training is being aimed at individuals who drive as part of their job, there may be a need for additional consideration of how best to tackle the training. Carlisle and Baden-Fuller (2004) and Gyi et al. (2013) have highlighted some of the problems encountered when attempting to provide training and information to workers who regularly drive as a normal part of their work. They highlighted the solitary nature of this population's work and stressed that group meetings were necessary for drivers if changes in attitudes were to be achieved. They had noted that during these meetings, drivers who were less convinced of the benefits of changing their working practices were influenced by more enthusiastic colleagues.

Training should occur away from the main work area so that the trainees can concentrate on what is being said by the course tutor, and so that they can rehearse practical skills without worrying about being observed by their work colleagues. A practical phase during training is essential as many trainees will never have used the specified techniques previously. The tutor should ensure that everyone is fit enough to do the practical exercises before inviting trainees to participate. Although it is always helpful to start the practical phase using small, compact loads, which provides trainees with an opportunity to get used to the basic principles of lifting without concern over heavy objects or irregular shapes, the session should also incorporate items representative of what the trainees would normally handle. This may need to include the use of trolleys, if that is an integral part of their work. Trainees should be encouraged to attend the course in loose-fitting trousers and flat shoes so that they can move easily. Women who generally wear skirts when handling loads will need some specific advice about the best way to approach a load if restricted by their clothing. Men should be advised to remove ties before starting the practical phase, and anyone with long hair should tie it back. Objects, such as keys, should be removed from trouser pockets so that they do not stick in the person's legs as they attempt to raise a load off the floor.

The first concept the trainees need to consider is planning and preparation. They need to understand the advantages to be gained from thinking ahead so that they can take steps in advance of the actual effort to reduce the risk—for instance, recognising that a door is closed and propping it open with a wedge before trying to pass through it carrying a load. Trainees should understand that they have to consider whether they have sufficient space to move around easily before handling a load. They should investigate the possibility of using a trolley rather than attempting to move the load without assistance. They should consider whether they need someone else's help and who might be the right person to help them. If they are faced with an unfamiliar load, the last thing they should do is attempt to pick it up. They should learn how to test the load initially to get an idea of its weight and distribution of weight before attempting to move it.

Having talked through the planning and preparation stage, the tutor can then move on to the actual handling phase of the course. They should first discuss and then demonstrate the handling techniques. Even though raising an object from the floor is an ideal starting point, handlers will not always pick loads up from the floor. However, the general principles employed in the basic floor lift are transferable to almost any load handled at any level. It should be kept in mind that using the basic floor lift is more demanding for the hips and knees than is the case when the handler simply bends forward at the waist to raise something off the floor. However, it is considered a better way to handle loads, particularly smaller compact loads, and to reduce overall risk.

The basic floor lift requires that the handler stands close to the load. This reduces the need to reach towards the load or to bend forward at the waist, both of which increase the risk of injury. The handler should stand with the feet hip width apart (see Figure 10.2). Trying to lift a load with the feet close together puts the handler in an unstable body position and makes them more likely to wobble as they bend their knees. Many handlers will also find that both heels come off the floor as they bend their knees, which will reduce stability still further. Once their feet are apart, they should step forward with one foot, placing one foot alongside the load if it is small enough to do so (see Figure 10.3). They will find themselves standing in a position similar to those suggested for contact sports, such as boxing, because it offers support and stability. Their feet should be in a 'ten and two' position rather than facing fully forward. Once the feet are in the correct position, handlers should bend their knees moderately to lower their hands to the point they wish to grasp on the load. If the feet have been turned out slightly as suggested, they will find that their knees will follow their feet and will not act as an obstacle between them and the load, which would happen if their feet faced forward. The handler's arms should remain within the boundary formed by their legs. They should not kneel on the floor nor over-bend the knees, because this will make standing up more difficult.

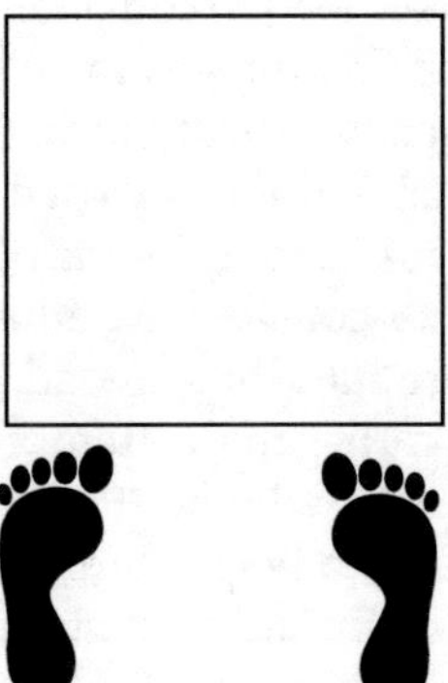

FIGURE 10.2
A handler standing with the feet hip width apart.

FIGURE 10.3
A handler standing with the feet hip width apart and one foot in front of the other.

The load should be grasped firmly using the whole hand. Leaning over the load slightly might make it easier to grasp; the handler does not need to be bolt upright, they just need to maintain a reasonably straight back. As handlers are likely to look down at the load as they grip it, they will find that the upper back will have bent forward in line with the inclined head. Just before standing up, the handler should raise their head and look forward. This will help to maintain an upright back. If they find that they have to jerk-lift the object to raise it from the floor, it is too heavy for them to move unaided. As the load is carried, it should be kept close to the body, and the heaviest side should be nearest the body. During the training session, once the load has been raised off the floor, the handlers can practice reversing the procedure and placing it back on the floor using the same techniques. If handlers have to work off a higher surface, they can still employ the same techniques of getting close to the load, feet apart, one foot in front of the other, upright posture, and good grip. They can bend their knees enough to bring them in line with the lower surface.

When pushing or pulling an object, handlers should still place their feet in the same basic position when starting to move the load. They should remain in an upright position, quite close to the load, and rely on their leg muscles to do the work. If they get stuck in a pot hole when using a trolley, they should avoid jerking it free. They should rock the trolley first to free it, or lean on the trolley with their full body weight to move it. Any abrupt movements should be avoided.

Before the training session, the trainer should be aware of what the handlers would move in a typical day, and objects representative of what they would move should be built into the practical session. The trainer should have determined in advance of the course whether they need to alter the demonstrated techniques to suit the characteristics of the loads. For instance, if handlers are going to be faced with raising large, bulky loads that would be difficult to pick up having bent the knees, the trainer should inform them that it is probably more important for them to keep the load close to the body than to bend the knees, which current research suggests is the least worst option in this situation. As a consequence, they may have to stoop. Also, female trainees who are likely to wear skirts when working should be offered advice on how to adapt the skills shown. For instance, wearing a skirt often limits the extent to which the female handler can position her feet. It also makes the knee bending posture a less dignified one. Female handlers should be advised how to work with their feet closer together but still with one foot as far in front of the other as the restrictions of their clothing will allow. They should also bend their knees as far as their clothing will allow, and they should maintain an upright head and upper body position as they lift the load. If they are severely limited by their clothing, they should seek assistance. They should not be required to handle anything large, heavy, or bulky if severely restricted by their clothing.

Once the trainees return to their work areas, efforts should be made to encourage them to apply their new skills. If they fail to employ appropriate handling techniques after their training, questions should be asked about why this is occurring. It may be that the trainees feel they can do it faster the old way; they may be embarrassed about their colleagues seeing them employ the new techniques; they may be rushing into their work without thinking; and some of them may find it hard to break old habits. The most effective way of encouraging the application of new skills is for managers and supervisors to lead by example and to reinforce the use of suitable handling techniques through praise.

The organisation should understand that providing manual handling training does not eliminate the risk associated with a badly designed manual handling task. For instance, if the trainee is required to pick up an excessively heavy load, it is unlikely that training will offer any significant protection. A review of task structure in combination with training would be advisable.

10.5 Summary

- Anyone who manually handles loads in an office environment should be given suitable manual handling training.
- The training should be tailored specifically to the type of handling they are likely to perform.
- Workers on the periphery of the main offices should be included in training (e.g. security staff, kitchen staff, mailroom staff, cleaners).
- It is not only heavy loads that cause problems. Light loads handled frequently, particularly using poor techniques, can also be problematic.
- Manual handling injuries can have quite a significant impact on a personal level and on a business level.
- The back is the part of the body most commonly injured during manual handling, but it is not the only body part that can be injured.
- Handlers can develop upper limb disorders as a result of repeated handling.
- Injuries to the back can involve a number of parts, such as the discs, ligaments, and muscles.
- Leaning forward at the waist, holding the load at a distance from the body, and twisting are prime causes of back injuries.
- Hernias can result from a combination of a weakness in the stomach wall, heavy lifting, and leaning forward as the load is lifted.
- Risk assessment is the most effective way of identifying the hazards associated with a manual handling task.
- Trolleys can lessen the burden of a manual handling task. They do not eliminate the manual element completely.
- The condition of a trolley affects the ease with which it can be moved. Damaged or worn wheels increase the resistance and effort involved in moving the trolley.
- Trolleys should be stacked properly to maintain stability.
- The wheels of a trolley should suit the weight of the load being moved and the surface across which it moves.
- All trolley wheels do not work equally well on all surfaces.
- The size of a trolley wheel and the number present have a bearing on how easy the trolley is to move.
- Whether the trolley wheels rotate at the front or rear influences how easy it is to steer.

- An individual moving a trolley on a slope should stand uphill of it.
- Trolleys regularly used on slopes should be fitted with brakes.
- Using vertical trolley handles allows individuals to choose a point to grip that suits them.
- Horizontal trolley handles are best for manoeuvring trolleys in confined spaces.
- Trolley handles with a diameter of 25–40 mm and covered with insulating material are recommended.
- There should be sufficient clearance for the handler's hands when they push a trolley through a doorway.
- Ideally, trolleys should be pushed in front of the body using two hands.
- The handler should be able to see over the trolley.
- The effort involved in starting the movement of the trolley and keeping it moving should be measured to ensure it is not excessive.
- Loads being carried should be held close to the body.
- Handlers should move their feet in the direction of movement to avoid twisting.
- Loads that are heavy or awkward should be presented between knuckle and elbow height.
- Moving loads through large vertical distances should be avoided. Loads should be placed on a waist-height halfway point if being moved from a high point to a low point or vice versa.
- Sudden movement of the load increases the risk of injury.
- The handler's ability to move loads safely is significantly reduced when they sit down.
- Carrying loads over distances in excess of 10 m increases the risk of injury.
- Two- and three-person lifts make the handling task more awkward.
- Large, bulky, or floppy loads increase the difficulty of the handling task.
- Handlers should be aware of the need to keep the heaviest side of an asymmetric load close to them.
- Packaging and handholds on a load influence how easy it is to grip.
- Steps and slopes make manual handling more demanding.
- Lack of clearance around the handler increases the risk of handling.
- Temperature and lighting levels are important considerations in terms of manual handling.

- Women have about two-thirds the strength capacity of men. It can diminish further during pregnancy.
- Age influences a handler's ability to cope with the demands of manual handling tasks.
- Businesses have ageing workforces, and they need to consider if their risk reduction measures need to remain fluid over time to match the changing needs of their workers.
- Unsuitable clothing and footwear can increase the risk of a handling operation.
- Training should include a practical session where trainees move loads typical of what they would handle at work.
- Trainees should learn the basic lift: feet hip width apart, one foot in front of the other, moderately bend the knees, straight back, head up, grip firmly, stand up smoothly.
- Managers, supervisors, and team leaders should lead by example and employ suitable handling techniques at all times.
- Specific, tailored manual handling training and advice needs to be given to any worker who uses a vehicle as part of their work.
- If rucksacks are used, they should be worn over both shoulders, chest and waist straps should be buckled up, the shoulder straps should be broad and padded to spread the load over a wider area, the rucksack needs to fit the length of the back properly, the material and frame of the rucksack should not make it heavy before it is filled, heavier items should be placed towards the bottom of the rucksack.
- Messenger bags worn over one shoulder should be avoided unless the items being carried are very light. The shoulder strap should be wide and padded and, preferably, the bag should be worn diagonally across the body.

References

Basri, B., and Griffin, M. J. 2012. Equivalent comfort contours for vertical seat vibration: Effect of vibration magnitude and backrest inclination. *Ergonomics* 55(8), 909–922.

Carlisle, Y., and Baden-Fuller, C. 2004. Re-applying beliefs: An analysis of change in the oil industry. *Organization Studies* 25(6), 987–1019.

Chen, J. A., Dickerson, C. R., Wells, R. P., and Laing, A. C. 2017. Older females in the workforce – the effects of age on psychophysical estimates of maximum acceptable lifting loads. *Ergonomics* 60(12), 1708–1717.

Deschenes, M. R. 2004. Effects of aging on muscle fibre type and size. *Sports Medicine* 34(12), 809–824.

Finch, J., and Robinson, M. 2003. Aging and late-onset disability: Addressing workplace accommodations. *Journal of Rehabilitation* 69(2), 38–42.

Gyi, D., Sang, K., and Haslam, C. 2013. Participatory ergonomics: Co-developing interventions to reduce the risk of musculoskeletal symptoms in business drivers. *Ergonomics* 56(1), 45–58.

Harris G., and Mayhog, L. 2003. Occupational health issues affecting the pharmaceutical sales force. *Occupational Medicine* 53(6), 378–383.

Loeser, R. F. 2013. Aging processes and the development of osteoarthritis. *Current Opinion in Rheumatology* 25, 108–113. doi:10.1097/BOR.0b013e32835a9428.

McCarthy, M. M., and Hannafin, J. A. 2014. The mature athlete. *Sports Health* 6(1), 41–48.

McKeown, C. 2011. *Ergonomics in Action: A Practical Guide for the Workplace.* Wigston, UK: IOSH.

Mitchell, W. K., Williams, J., Atherton, P., Larvin, M., Lund, J., and Narici, M. 2012. Sarcopenia, dynapenia, and the impact of advancing age on human skeletal muscle size and strength; a quantitative review. *Frontiers in Physiology* 3(260), 1–18.

Porter, J. M., and Gyi, D. E. 2002. The prevalence of musculoskeletal troubles among car drivers. *Occupational Medicine* 52(1), 4–12.

Schleifenbaum, S., Prietzel, T., Hädrich, C., Möbius, R., Sichting, F., and Hammer, N. 2016. Tensile properties of the hip joint ligaments are largely variable and age-dependent – An in-vitro analysis in an age range of 14–93 Years. *Journal of Biomechanics* 49(14), 3437–3443.

11

Work-Related Ill Health

11.1 Introduction

Office workers complain commonly of a number of health problems, such as musculoskeletal disorders, which include upper limb disorders (ULDs) and backache, as well as eye fatigue, headaches, and stress. Musculoskeletal disorders (MSDs) are widespread throughout the world, and are associated with enormous financial and societal costs (Gallagher and Schall 2017).

Some individuals believe that using computers virtually guarantees that they will experience some form of ill health. That is not necessarily the case. Although an association between computer work and MSDs of the neck and upper limbs has been established (Gerr et al. 2006, Levanon et al. 2012), the important factor is not whether people use this equipment, but under what circumstances they use it. It is worth stressing that often it is very simple aspects of the working environment that create quite significant pain and discomfort. However, the positive element of this is that these simple aspects can be addressed quite easily and can, in almost all cases, be rectified. Even if it is not possible to eliminate a 'provoking' factor identified as the cause or trigger of the difficulties experienced by a worker, it is possible to control the impact of the adverse working conditions by limiting an individual's exposure to them.

11.2 Upper Limb Disorders

Upper limb disorders (ULDs), also called cumulative trauma disorders, is an umbrella term that indicates an injury has occurred anywhere from the fingertips, along the arm, and into the shoulder, and, possibly, the back of the neck. These disorders are not suffered only by individuals performing computer-based operations; they can be experienced by any worker who is exposed to the right type of provoking factors under the right circumstances. ULDs are part of a broader category referred to as musculoskeletal disorders. They are soft tissue conditions that can affect muscles, tendons, ligaments, and joints. All of these body parts work together to produce a system that is responsible for varying limb positions and limb activity. The activity can either be static or dynamic. Static muscle activity occurs when the limb or digit is held in a fixed position, unsupported by any surface. Dynamic muscle effort is responsible for movement. Limb movement relies on a complex combination of static and dynamic work on the part of the muscles.

Limbs are considered to work most efficiently when moving within a natural/neutral, or comfortable, range. Over-extension of a joint, or the adoption of an irregular posture, so that the joint moves beyond its natural range, is more likely to result in the development of a ULD if performed repeatedly or is sustained for an extended period. Irregular wrist positions include ulnar and radial deviation and extension and flexion of the wrist. Ulnar deviation relates to the bending of the hand at the wrist in the direction of the little finger. Radial deviation relates to the bending of the hand at the wrist in the direction of the thumb. Extension relates to bending the hand upward at the wrist and flexion relates to bending the hand downward at the wrist. These positions are illustrated in Figure 11.1.

Other, more gross movements of the limbs are also associated with the development of ULDs, particularly if carried out repetitively, forcefully, or are sustained for extended periods. These include supination, the rotation of the forearm to bring the palm of the hand facing up; pronation, rotation of the forearm to bring the palm of the hand facing down; abduction, moving the arm outward and away from the body; and adduction, moving the arm across the midline of the body (see Figure 11.1). Generally, everyone adopts these postures during the course of the day while talking to others, performing tasks, and so on, and they experience no adverse consequences. The difficulty arises when individuals use these irregular postures on a frequent basis as a standard part of their work and do so for extended, uninterrupted periods.

Some individuals experience symptoms in only one place, such as at the wrist, while others encounter symptoms in several places. The sites at which the symptoms are present and the type of signs and symptoms displayed allow a specific diagnosis to be made. Typical symptoms and signs include pain, swelling, tenderness, aching (often described as similar to toothache), and underlying discomfort. Sufferers may also experience sensations such as tingling, 'pins and needles', a feeling of heat, or a more extreme burning sensation, numbness, or sensations described as electric shocks. They may also find that their hands have become stiff, or weaker, making it more difficult to carry objects. The grip might not be as strong as it once was, making it difficult to perform certain tasks, such as opening a jam jar or bottle cap. They may find that, generally, their hands are less dexterous, and they may occasionally experience cramps

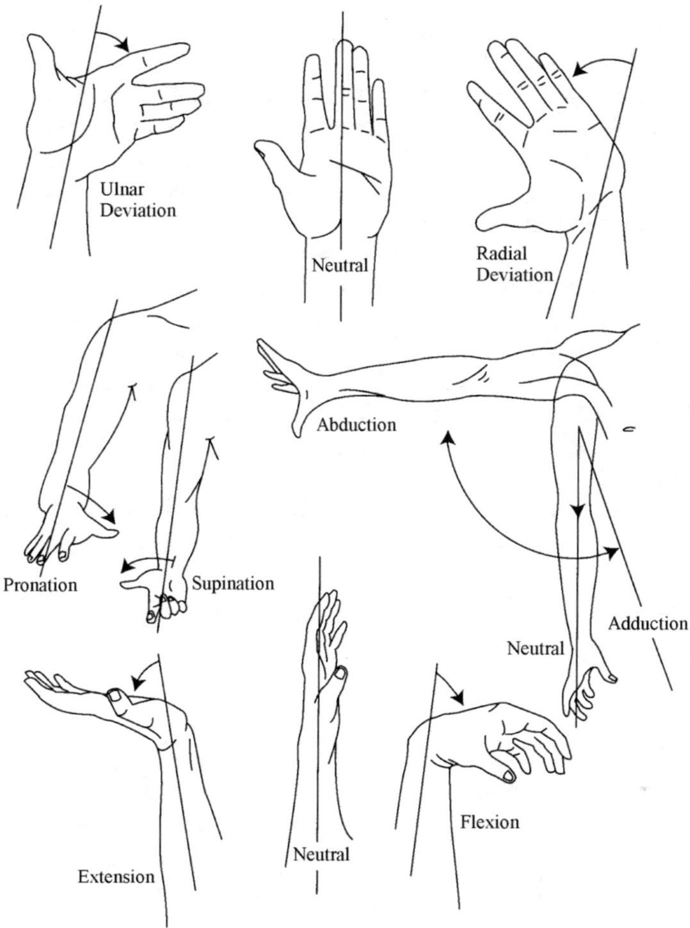

FIGURE 11.1
Irregular wrist and arm positions alongside the neutral position of the hand and wrist. (From Putz-Anderson, V., *Cumulative Trauma Disorders: A Manual for Musculoskeletal Diseases of the Upper Limbs*, Taylor & Francis Group, London, UK, 1992.)

or spasms. An observer may notice that parts of the sufferer's limbs are swollen, or there may be a change in skin colour. They may hear a crackling sound when some areas of the limb, such as the wrist, are moved, and they may notice a change in the ability of the sufferer to grip, move, or operate objects. The sufferers of these conditions will find that the pain or discomfort is not transitory, in contrast to the general aches and pains that many workers experience at some time in their working life. The sooner individuals are assessed, given a diagnosis and treated, the more likely it is that they will make a successful recovery. Without appropriate intervention, some individuals may start to experience severe and chronic pain and impairment of limb function. Ultimately, they may be left with residual disability.

It is possible for an organisation to get an indication of the level of discomfort experienced by their workforce by carrying out a discomfort survey. Figure 11.2 is a commonly used body parts discomfort questionnaire, probably based on the original concept of Corlett and Bishop (1976). It is given to the workers who indicate where they are experiencing discomfort by shading appropriate areas of the body. This allows the organisation to identify whether their workers are free of discomfort, or whether there are trends of discomfort that might be related to specific types of work or specific departments. This is important, as Dennerlein and Johnson (2006) reported that different computer tasks have different levels of biomechanical risk. This type of questionnaire is considered (e.g. Van Eerd et al. 2009) an inexpensive, relatively low burden method for gathering task information on large numbers of workers.

Employees can also be asked to identify on this body parts discomfort questionnaire what type of symptoms they are experiencing, such as pins

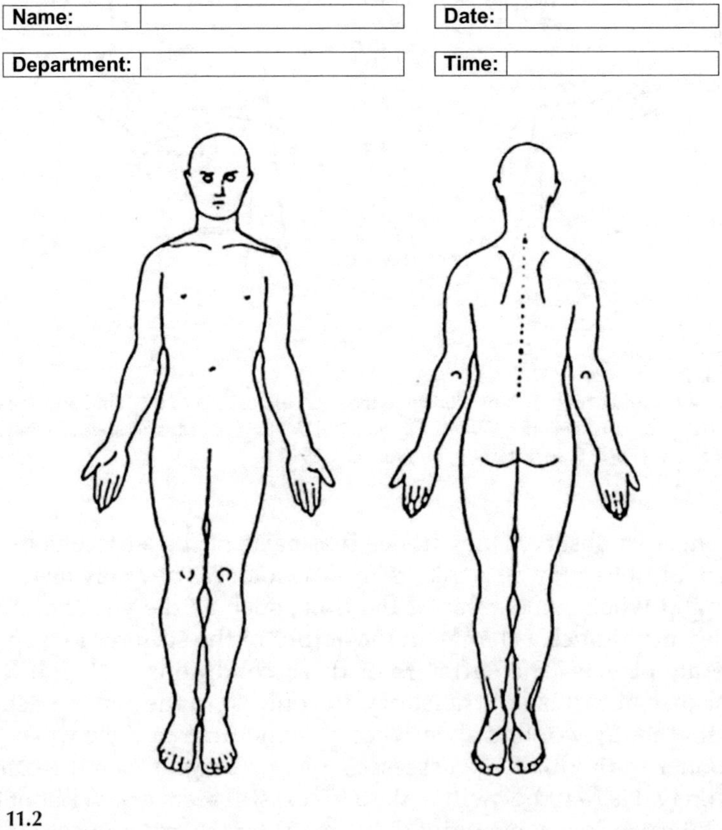

Name:		Date:	
Department:		Time:	

FIGURE 11.2
Body parts discomfort questionnaire.

and needles or aching, and they can be asked to rate the extent of their discomfort on a scale of one to five by writing the number alongside the area of concern. As it is common for symptoms to present themselves and to change at different times of the day, using the questionnaire shown in Figure 11.3 will help to plot what occurs during the course of a working day. Workers should be advised, and reminded, that the questionnaire should be completed at predetermined intervals throughout their shift.

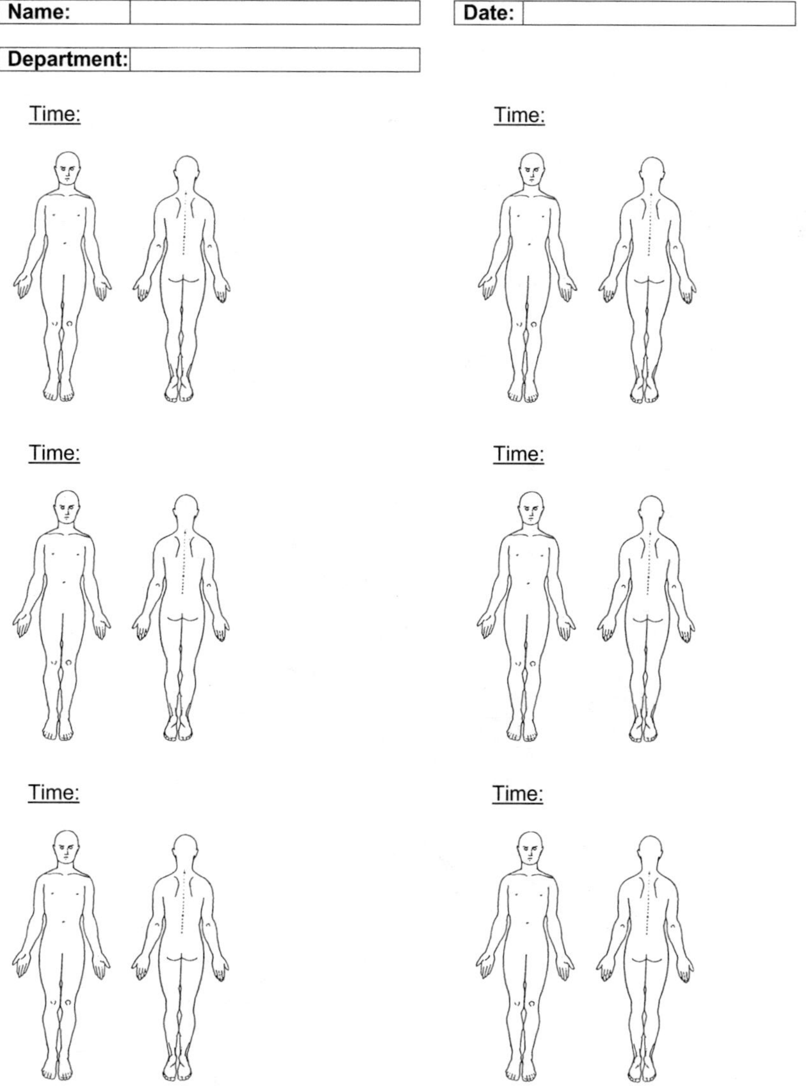

FIGURE 11.3
Body parts discomfort questionnaire for completion at intervals throughout the working day.

11.2.1 Types of Upper Limb Disorders

There are a number of disorders that fall under the heading of ULDs. For some time, a few of them were used erroneously as catch-alls for any ULD in any location. For instance, tenosynovitis and carpal tunnel syndrome were mistakenly used as general terms to denote that an individual had an upper limb disorder, but, in fact, these are very specific disorders located in specific sites of the upper limbs. They are diagnosed on the basis of the location of the anatomical structure affected by symptoms.

11.2.1.1 Tenosynovitis

This condition is centred in the wrist area and relates to an inflammation in the lining of the synovial sheath in which a tendon is enclosed. Figure 11.4 shows the tendons in the hand and the synovial sheaths. Tendons are the means by which muscle attaches to bone and across which the muscle transmits forces that result in movement of the individual bone and the limb as a whole. Under normal circumstances, the tendons can move smoothly within

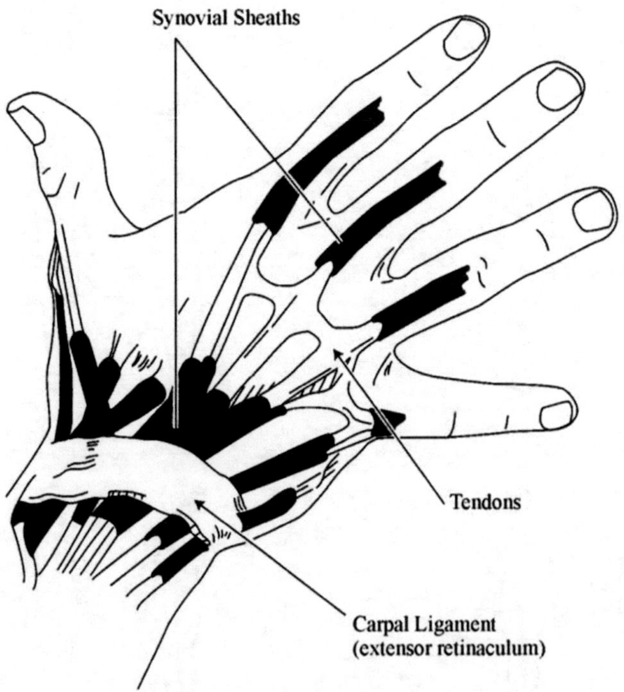

FIGURE 11.4
Illustration of the hand showing the tendons and synovial sheaths. (From Putz-Anderson, V., *Cumulative Trauma Disorders: A Manual for Musculoskeletal Diseases of the Upper Limbs*, Taylor & Francis Group, London, UK, 1992.)

the sheath, and this is assisted by the presence of some lubrication. The sheaths are intended to offer protection to the tendons as they pass under ligaments or as the wrist is bent and the tendon is pressed against the bones of the wrist.

Work that is highly repetitive, forceful and/or requires the adoption of irregular wrist positions, as shown in Figure 11.1, is closely associated with the development of this condition. Traumatic injury, such as injuring a wrist during a bad fall, may predispose the individual to developing tenosynovitis.

The individual suffering from this condition is likely to experience aching in the wrist area, which is likely to worsen with use of the wrist. It is usually accompanied by local tenderness and swelling and, sometimes, weakness in the hand, often displayed by deterioration in the ability to grip and hold objects. The sufferer is sometimes aware of an audible creaking sound when moving the wrist, often described as sounding like someone walking over dry snow. This is known as crepitus.

If the movement of the tendon within the sheath is restricted, the condition is referred to as stenosing tenosynovitis. The sufferer may be aware of a clicking or pulling sensation when trying to extend the fingers or thumb. This condition is often associated with overuse of the wrist during work activities. De Quervain's tenosynovitis involves the abductor pollicis longus and extensor pollicis brevis tendons, which form the 'anatomical snuffbox', or small indent, clearly visible at the base of the thumb. Prolonged or repeated effort and unaccustomed work have been associated with the development of this condition, particularly when forceful gripping is combined with deviation of the wrist. Heavy reliance on the use of the thumb when completing an activity is also highlighted as precipitating the onset of this disorder, more so if the gripping action requires the thumb to be moved away from the main body of the hand, such as when spanning a large object with the fingers and thumb or using a tool like scissors. If the tendons of the finger flexors are compromised through overuse, the individual will be diagnosed as suffering from trigger finger. The finger flexors are responsible for closing the fingers to bring the fingertips in contact with the palm. If the tendon sheath becomes swollen, the tendon will not be able to move smoothly and will only be able to make jerky movements. This condition is often associated with overuse of the fingers and with repetitive or extended gripping of an object with a hard or sharp edge.

11.2.1.2 Carpal Tunnel Syndrome

The carpal tunnel is formed by eight small carpal bones in the wrist and the flexor retinaculum (also referred to as the carpal ligament). The finger flexor tendons travel from the forearm and pass under the carpal ligament and through this tunnel inside their synovial sheaths alongside the median

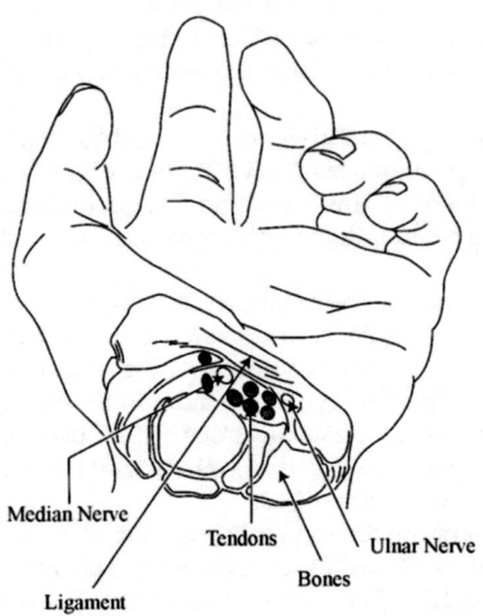

FIGURE 11.5
Illustration showing the location of the median nerve. (From Putz-Anderson, V. *Cumulative Trauma Disorders: A Manual for Musculoskeletal Diseases of the Upper Limbs*, Taylor & Francis Group, London, UK, 1992.)

nerve before inserting into the fingers (see Figure 11.5). The median nerve provides sensation to the thumb, middle finger, index finger, part of the ring finger, and a major proportion of the palm.

Carpal tunnel syndrome results from compression of the median nerve. This can occur if there is swelling and irritation within the confines of the carpal tunnel, which can be a consequence of tenosynovitis. Forceful gripping, irregular wrist postures, and vibration are associated with the development of carpal tunnel syndrome. It has also been connected with rapid finger movements (Bridger 2017). Typical symptoms include numbness and tingling, especially in the areas of the hand served by the median nerve. It is not uncommon for sufferers to wake during the night experiencing these symptoms. The hand may experience a loss of sensation and weakness, which may make the sufferer appear to be more clumsy than usual. Pregnant women and women who use oral contraceptives are also known to be predisposed to this condition, and it can coexist with other disorders such as diabetes and arthritis.

Although the aetiologies of disorders, such as carpal tunnel syndrome, are multifactorial; it is believed (e.g. Donoghue et al. 2013) that computer users are at a high risk of suffering from this particular condition. Aydeniz and Gursoy (2008) and Conlon and Rempel (2005) would suggest that there is an increased risk of carpal tunnel syndrome with increased computer use.

11.2.1.3 Dupuytren's Contracture

Dupuytren's contracture results from a thickening of the fibrous tissue on the palm of the hand, referred to as the palmar fascia. As the thickening progresses, it is accompanied by bending of the ring and little fingers, usually more prominent in the little finger. This is usually permanent, unless there is surgical intervention. The condition is painless and tends to be more inconvenient than anything else, because the sufferer may find that their fingers become a hindrance as they try to put their hands in their pockets or if they try to wear gloves. Although this condition can be congenital and therefore not work-related, an association has been established with repeated minor trauma to the hand, such as if banging envelopes in a sorting machine flat with the palm, specific injury, such as cutting the palm with a knife or scissors, and the use of vibrating tools.

11.2.1.4 Vibration White Finger

As its name suggests, vibration white finger (VWF) is associated with the use of vibrating equipment. Although it might be assumed that this is unlikely to occur within an office environment, some commercial organisations use machinery that vibrates, such as sorting machines or letter opening machines. The disorder is caused by an impairment of blood circulation in the fingers. The symptoms typical of VWF include blanching of the fingers, and this is usually accompanied by numbness, tingling and the fingers feeling cold. Once the blood flow is restored, the fingers become red, throbbing, and painful, after initially appearing bluish. The blue or purple colour may become more long-lasting if the condition worsens. As time progresses, more of the hand becomes involved and the attacks become more frequent. Coordination can be impaired, as can dexterity. In some very rare and extreme cases, VWF has led to gangrene in the digits.

Some office workers may suffer from another condition that results in similar symptoms, such as a change in colour in the hands and pain in the fingers. This is Raynaud's phenomenon, a constitutional disorder usually experienced in both hands. The symptoms are most often provoked in cold conditions, particularly in winter.

11.2.1.5 Ganglion

A ganglion is a fluid-filled cyst, usually found around a joint or tendon sheath, often on the back of the hand or wrist. The fluid is synovial fluid, and the cyst is usually about the size of a pea, but can get bigger. Although there is no generally accepted association between working conditions and the appearance of ganglion, most ergonomists would probably agree that their presence in large numbers amongst a workforce suggests that the workers are exposed to inappropriate working conditions.

11.2.1.6 Epicondylitis

A more everyday term for epicondylitis is tennis elbow, or golfer's elbow, depending on which side of the elbow is affected. Lateral epicondylitis (tennis elbow) affects the side of the elbow that lies directly in line with the thumb when the palm of the hand is facing upward. It occurs when the epicondyle, which is the point of origin for the wrist and finger extensors, becomes inflamed. Overuse of the muscles responsible for extending the hand and fingers may cause pain at their point of origin, and this pain can radiate down the arm. It is generally accepted that repeated forceful movements of the wrist and forearm, such as wrist extension and rotation of the forearm, are likely to contribute to the development of this condition. Forceful throwing and heavy lifting, particularly when the back of the hand is facing up, are also thought to be linked with this condition.

Medial epicondylitis, or golfer's elbow, is not as common. It results from irritation at the point of origin of the flexor muscles, which bend the hand downward at the wrist, on the inside of the elbow and is usually associated with repeated flexing of the wrist and rotation of the forearm.

11.2.1.7 Tendinitis

Tendinitis can affect any tendon in the body. Depending on the tendon affected, movement can become limited. Humeral or rotator cuff tendinitis affects the tendons associated with the muscles that rotate the arm at the shoulder and move the arm away from the body. The sufferer experiences aching around the shoulder, which can be triggered by lying on the affected shoulder during sleep. Forceful work and repetitive operations have been associated with the development of this condition (Violante et al. 2000), as has working with the arms raised in excess of 60° above the shoulder.

11.2.1.8 Frozen Shoulder

Sufferers from frozen shoulder start by experiencing a gradual onset of stiffness, accompanied by shoulder pain that is more prominent at night. Ultimately, they experience significant restriction in all shoulder movements. This results from the inflammation in or degeneration of the shoulder joint tissue. Overhead work that is repeated or sustained for extended periods is believed to contribute to this condition. It can also occur spontaneously without any identifiable provocation.

11.2.1.9 Thoracic Outlet Syndrome

Thoracic outlet syndrome occurs when the nerves and blood vessels between the neck and the shoulder are compressed. The sufferer experiences symptoms, such as pain in the underside of the arm and numbness in the fingers.

Working with the arms raised or working with the shoulders pulled back and downward, such as carrying a suitcase, are associated with the development of this condition. It is believed that repeatedly bearing loads directly on the shoulders or suspending weights from the shoulders, such as carrying a laptop in a case using a shoulder strap, will increase the likelihood of developing this condition.

11.2.1.10 Cervical Spondylosis

Cervical spondylosis is a condition affecting the neck and spine. It results from degeneration of the discs or intervertebral joints, which can irritate the nerve roots in the cervical spine. The sufferer experiences pain in the neck and shoulder. It has been associated with age, but also with tasks that require people to carry heavy loads on their shoulders.

11.2.1.11 Osteoarthritis

The more accurate term for osteoarthritis is 'osteoarthrosis'. It is a degenerative condition affecting the articular cartilages of the synovial joints. It occurs in conjunction with changes in surrounding bone and can be sited at any articulating joint. It is more likely to occur in a joint that has been subject to long-term stress. Typical symptoms include stiffness, aching on movement, and restriction in movement. Pain can radiate from the neck into the arms. Sometimes a grating sound can be heard on moving the joint.

11.2.2 Causes of Upper Limb Disorders

Although it is generally accepted that workplace factors can cause or contribute to the development of ULDs, non-work-related factors can also be the source of these conditions. Non-work-related factors can also aggravate the conditions once they have surfaced. For that reason, once an individual reports experiencing symptoms that suggest a ULD, enquiries need to be made about life outside of work as opposed to focussing solely on the activities performed in work. However, having accepted that non-work-related factors can be connected with the development of ULDs in some circumstances, it has to be recognised that it is unusual for people to spend as much time on leisure pursuits as they do on work activities. In addition, when someone is playing the piano or sewing or carrying out a do-it-yourself task at home, they can stop whenever they feel the need for a break, for as long as they wish, which is in stark contrast to working environments, where people usually have to continue working until they are permitted to take a scheduled rest break of predetermined length.

When considering the causes of ULDs, we can divide them into two main categories: the main causative factors and the contributory factors. The main causative factors are repetition, awkward posture, static muscle work, force, and duration of the activity. Exposure to any or all of these factors does not guarantee that an individual will develop a ULD. Usually, the causative factors need to be combined with other contributory factors, such as lack of rest breaks, lack of task variety, and extended working periods, to result in the development of a disorder.

Recently, various authors (e.g. Gallagher and Heberger 2013, Gallagher and Schall 2017) have started to consider if there is a fatigue failure process in musculoskeletal tissues. This has grown from the knowledge that all materials experience failure through either: application of a one-cycle high-magnitude stress at the 'ultimate stress' point of the material or (2) repeated application of loads that are a percentage of the material's ultimate stress point. Gallaher and Schall (2017) believe that the implications of an underlying fatigue failure process in MSDs development have generally not been considered in prior MSDs risk assessment tools or MSD prevention strategies. They feel it is important to develop improved cumulative loading estimates on tissues.

Coincidentally, wearable protection technology has become much more common in recent years. Their application in the workplace has been helped by the fact that people are completely comfortable with using domestic fitness trackers and are open to accepting other technologies that can be used to manage their health, safety, and fitness. Recently, specific body-worn sensors have been developed that allow data to be collected whilst the individual is at work and analysed in an effort to identify repetitive movements and poor posture. It is hoped that this will provide a sound basis for redesigning the task if it is highlighted as being problematic.

11.2.2.1 Repetition

When individuals perform a task, they rely on their muscles to move, support, and stabilise them. If the work involves the same sequence of tasks repeated in rapid succession over an extended period, the task can be referred to as repetitive. If the task is repetitive, then the muscle groups responsible for assisting the individual to complete the task will be used in the same manner repeatedly. If the work is rapid or repeated over an extended period without interruption, the muscle groups will not be able to recover fully from the work and will be subject to fatigue. Repeatedly overloading the muscle groups in this way can lead to inflammation and degeneration of the soft tissue. If people are working at a very rapid rate or performing work where they have to accelerate their movements, they will be employing high muscle forces. The application of force can also increase the rate at which the muscles fatigue.

Some work is carried out at such a rapid rate that it is categorised as highly repetitive. A task can be identified, using Putz-Anderson's (1992) classification, as highly repetitive if it has a cycle time of 30 seconds or less or if 50% of the cycle involves the same kind of fundamental cycle. An example of the latter can be found in a company whose business involves the packing of book orders. It might take an employee one minute to load selected books into a box ready for dispatch, which might initially suggest that the task is not highly repetitive. However, they might place 10 books in the box at a rate of one book every six seconds (i.e. the fundamental cycle is repeated every six seconds), which would result in this task being considered highly repetitive. Individuals who perform highly repetitive tasks are at greater risk of developing a ULD.

Keyboard and mouse work typically involve repetitive operations, as keys are depressed and as buttons on the mouse are clicked. Users tend to operate keyboards and mice in a stereotyped manner over extended periods. Repetitive movements during computer work have been identified as one of the risk factors for development of musculoskeletal disorders (Karlqvist 2002). In some operations, such as data-entry tasks, users are encouraged to work at an even faster rate by being offered financial incentives for inputting larger volumes of data. Many keyboard users depress keys at a rate that does not reach 10,000 depressions per hour, which is considered to be the point at which keyboard operators move into a higher risk category. Data-entry operators, when offered incentives, can key in at rates ranging from about 16,000 to 25,000 depressions per hour, which, unsurprisingly, moves them into the higher risk category.

The repetitive nature of keyboard and mouse work can be altered only through changing the speed of the operation and through changing the structure of the work so that there is a greater variety of tasks. If these changes cannot be implemented, then the most effective way of combatting the effects of performing a repetitive task is to use rest breaks. As a general rule, short frequent breaks are more effective at combatting fatigue than fewer longer breaks. On that basis, a five-minute break every hour is likely to be more effective than a longer break after two or three hours of work. Such a schedule should be offered on a consistent basis, day to day, and not just when it suits the work demands at the time. Whenever individuals take a break, it would be in the company's best interest to ensure that they do not employ this time to use their computer system for personal reasons, such as checking personal emails or looking at material on the Internet. The purpose of the break is to ensure that their screen-based work is interrupted. Arvidsson et al. (2006) showed that although air traffic controllers did not have to use their computer system for one-third of their working shift, many of them tended to use their computers for personal reasons during their rest times. Therefore, although it would appear that they had significant periods of time when they did not need to operate a computer system, which sounds ideal, they actually used the system longer than was necessary owing to their own personal requirements, which undermined the benefits of the intended non-screen time.

11.2.2.2 Awkward Postures

The postures adopted by individuals when sitting at a desk and when operating a keyboard and mouse play a significant part in determining whether they are likely to encounter physical problems while they work. Constrained postures have been specifically identified as risk factors in the development of ULDs (Gerr et al. 2004, Karlqvist 2002). This underlines the importance of ensuring that workstation furniture is used correctly, and that the equipment on the work surface is laid out in a suitable arrangement. People who work in a neutral position are less likely to encounter a ULD. An awkward posture is one in which a limb (or other part of the body) is used in a position outside its optimal range. For instance, locating a mouse or keyboard too far forward on the desk will result in the user having to extend their arm forward, and reaching in this manner results in an awkward posture. Computer users often bend the hands up at the wrist as they lean their wrists on the desk surface (or wrist rest) and continue to use the mouse and keyboard from that position. Some users position the mouse out to the side so that they have to bend their hands at the wrist in the direction of the little finger. Other users set their chairs so low relative to the desk that they have to raise their shoulders and arms when operating the mouse and keyboard. Maintaining these postures requires additional muscular effort because the muscles cannot work as efficiently when the limb is working at the extremes of its range. This can result in injury owing to the friction and compression of soft tissue.

11.2.2.3 Static Muscle Work

Static muscle work is used to hold a limb, or any other body part, in a fixed position for an extended, uninterrupted period. During this time, the muscle is tensed and unable to relax. Blood flow to the soft tissue is restricted, and there is a build-up of metabolic waste. Muscles used in this manner tend to fatigue at a rapid rate. Some individuals may find this surprising because 'static' muscle work implies that no movement is taking place, and they may assume that when no movement occurs the limb must be resting. It is very easy during the assessment process to overlook static muscle work and, thus, its contribution to the development of a ULD, because the eye is not naturally drawn to non-movement. It tends to be drawn to movement, which results in work involving movement often becoming the focus for the assessment.

It takes 12 times the length of the actual activity requiring static muscle work for the muscle groups involved to recover fully. In other words, if a keyboard operator sits too low relative to their mouse and keyboard, and raises their shoulders and arms so they can use this equipment effectively, and maintains this position for five minutes, it will take the muscle groups involved in generating the required posture about one hour to recover fully.

In the case of mouse use, users may click the left button of the mouse with their index finger (i.e. employing dynamic muscle work) whilst holding their remaining fingers above the mouse to avoid inadvertent activation (i.e. employing static muscle work). Lee et al. (2008) have reported that the sustained, static muscular activation patterns of the finger extensor muscles required to lift the fingers during mouse use may play a role in the occurrence of forearm and hand and wrist pain during intensive mouse use.

Some studies (e.g. Onyebeke et al. 2014) have found that participants reported less musculoskeletal discomfort when using a support for their wrist or forearm when operating a mouse. However, this would have no impact on the problems associated with raising the fingers to avoid depression of the mouse buttons. The benefits were aimed at reducing shoulder muscle activity and torque by supporting the arm, which was suspended in a fixed position above the desk and mouse during mouse operation.

Sustained static posture has been identified as one of the most important risk factors for work-related neck and upper limb disorders in computer users (Yang and Cho 2012). This issue has quite serious implications in situations where an individual might employ static muscle work, as in this example, for the whole working day. Gerr et al. (2002) identified static and awkward positions as significant risk factors in the development of ULDs. Apart from ensuring that appropriate postures are adopted from the outset of the working day, it is also important that people get regular opportunities to leave their workstations so that they relieve the stresses imposed by remaining in a fixed position.

11.2.2.4 Force

The level of force generated by individuals is influenced by the postures they adopt. If they work in irregular or awkward postures, they will use increased muscular effort. Increased effort is also required if they have to work quickly. Engaging high levels of force is fatiguing, and if it is a consistent feature of the work, it can lead to injury. Apart from the fact that many computer operators adopt irregular postures when working and depress keys at a rapid rate, many of them use excessive force when depressing the keys. People who have received specific keyboard training, such as traditional secretaries, are probably in the minority in most businesses, because many people learn how to operate a keyboard by watching others or just having a go themselves. As a consequence, they do not acquire specific keyboard skills. Keyboards are designed to be touch-sensitive, and the keys can be depressed with minimal effort. Yet, untrained keyboard operators frequently hit the keys with excessive force, which will do nothing but increase the workload for the muscles concerned. This is an issue that should be tackled through training, along with advising workers how they should use a mouse. Many people grip the mouse with more force than is required, and, as a result, they increase the rate at which their muscles fatigue.

Forces are also applied during the manual handling of loads. Manually handling loads can result in the development of ULDs, given the right circumstances. As the loads increase in size and weight, more force will be required to lift them or push and pull them. Reducing load size and weight will reduce the effort required to move them. Reviewing the packaging used to contain a load is advisable, as this also influences the degree of force employed when moving it. If the load is contained in a large box with no handholds, it will require more force to carry it than would be the case if it sat in a tub with handholds on the side.

11.2.2.5 Duration of Exposure

When considering the issue of duration of exposure, a review is made of not only how many hours per day an individual might work, but also how many days per week they work. Recent studies of computer work have shown that the daily duration of computer work was positively related to the risk of pain and discomfort (Juul-Kristensen et al. 2004). The longer people use a computer, the more likely they are to develop a ULD (Griffiths et al. 2007, Gerr et al. 2002). This probably results from the fact that many ULDs are considered to be cumulative in nature, and the longer an individual spends performing a task that is hazardous in some way, the more likely they are to encounter difficulties. Working for longer periods is always accompanied by shorter periods of time being available for rest and recovery. This has implications whenever a business relies on employees working overtime.

The discussion of duration of exposure does not imply that minimal exposure time will not result in the development of a ULD. If the task is extreme enough, a short exposure time might be sufficient to result in the appearance of symptoms. For instance, requiring a secretary to stand in for an absent data-entry operator on one or two occasions may be sufficient to trigger the onset of a disorder, given the nature of data entry and the fact that the secretary has not been given a period in which to adjust to the demands of the job. This is also true for people returning to work after an extended leave. Attempting to pick up where they left off prior to their leave may result in them developing symptoms of a disorder within a very short time.

11.2.2.6 Other Contributory Factors

The importance of rest breaks and changes in activity has been dealt with several times throughout this book. In particular, their importance must be emphasised in relation to avoiding ULDs. People who repeat the same type of activity for extended periods without interruption are more susceptible to developing a ULD than those who have variety in their work and who are able to take regular short breaks. Maintaining a consistent pattern of work and minimising peaks and troughs is also recommended so that workers are not exposed to conditions that exceed their level of task fitness, making them more susceptible to injury.

Lack of training and suitable supervision can also contribute to the development of ULDs. Workers cannot be assumed to know how to work safely, unless someone has taken the trouble to point out the risks and how to avoid them. Once they have received the training that allows them to do their work safely, they need to be monitored by competent supervisors to ensure that they employ appropriate working practices at all times, including periods when they are under pressure to get an assignment completed.

Working in cold environments or being in contact with cold surfaces is known to increase the possibility of an individual developing a ULD. Although it is unlikely that office workers will work in an environment with low temperatures, they may come in contact with cold surfaces. This can occur if their desks are constructed of materials that do not warm up during the day. For example, placing aluminum edging on desks can result in workers' forearms coming in contact with the cold surface and heat exchange occurring. Some people work in very old offices with poor heating and draughty windows, which results in localised cold spots in the office. Working consistently in this type of environment could increase the likelihood of them developing a ULD if they perform a task that is inherently hazardous.

Finally, vibration is a key contributor in the development of ULDs. However, few people work in an office environment with vibrating equipment or who are otherwise exposed to vibration. Nonetheless, some commercial businesses use large pieces of equipment at their sites, for instance, to open envelopes, to process cheques or forms, or for certain reprographics processes. All of these can be sources of vibration.

11.2.2.7 Psychosocial Factors

Psychosocial factors have been strongly associated with upper limb pain (e.g. Cagnie et al. 2007). The context in which work is done and the content of the work have an impact on the individual's psychological response to it. The context of work relates to the social aspect of the workplace and the relationships prevalent in the environment between workers on the same level within the hierarchy and between different levels of the hierarchy. The content of the work relates to matters, such as the control individuals feel they have over how they do their work, whether it is challenging enough or too challenging, whether they feel they can apply all of the skills they have developed, how much variety they have in their work, and so on. It is believed that an individual's psychological response to their working conditions will be significant in determining whether they develop ULDs when exposed to physical working conditions that are problematic. To limit the likelihood of someone developing a ULD, it is necessary to focus not only on physiological aspects, such as posture and speed of operation, but also on psychosocial elements.

11.2.2.8 Extra-Organisational Factors

There are certain leisure pursuits and hobbies that could be identified as triggers in the development of ULDs, assuming they are performed frequently enough and for periods of sufficient length. Playing musical instruments, knitting, bell-ringing, computer games, and, more recently, texting on mobile phones have been implicated in the development of these conditions. Racquet sports, in particular, are highlighted as being problematic. Given the number of people who visit the gym on a regular basis and their propensity to work out without guidance and ongoing monitoring from qualified, informed staff, there must be a question mark over whether any upper limb injury could actually be caused by or aggravated by inappropriate practices at the gym. There are also personal factors such as previous injuries and illnesses that can predispose an individual to developing a disorder.

11.3 Responding to ULDs

The best response to complaints suggestive of a ULD is an immediate one. As ULDs tend to get progressively worse, the sooner action is taken, the more likely it is that the individual will recover successfully. To assist the rapid response, it is important that people who are experiencing symptoms recognise their implications and realise that they need to report them formally to their manager or supervisor so that action can be taken. This can only occur if workers have been warned about the risks associated with their work and the symptoms they may encounter. It is also only likely to occur if sufferers feel comfortable about starting the ball rolling by reporting they are having problems. If they feel intimidated in any way or fear the loss of their jobs or potential for promotion, they may not report their symptoms, thereby allowing the condition to worsen. Once they report their difficulties, their point of contact must appreciate the urgency of what they are being told and must respond accordingly. To that end, supervisors also need the right type of training and awareness.

It would benefit the managers to have a clear diagnosis, as soon as possible, so they know exactly what action to take. This may require referral to a specialist rather than relying on the sufferer's own doctor to provide an accurate diagnosis and regime of treatment during the allotted five- or ten-minute appointment slot at their local clinic. There may be no connection between the condition and the workplace. However, if there is, it would be useful for the organisation to be advised by the physician on how to respond.

For example, organisations need to know whether the injured party can continue to work, and if so, whether they can continue with it in the present form. They need to know if the sufferer should work only for short periods, or perhaps not be working at all. The difficulty of seeking advice from the injured party's own doctor is that the doctor will have based the diagnosis and subsequent advice on the work history provided by the employee. Doctors may misinterpret what they are told, especially if they have had no hands-on experience of that type of environment, or employees may provide a rather biased view of the work they perform; and on that basis, the doctor may suggest an overly cautious regime of rehabilitation. Any physician being asked to provide advice on how best to treat injured workers should be provided with an accurate account of the work they perform. In other words, the business should state in clear terms exactly what the individual's work entails and specifically ask the physician whether there are aspects of it that should be limited or avoided.

It is not uncommon for an individual who is suffering from a ULD to be given leave from work for a few weeks or months to rest the limb. Once people return to work, they should not be expected to pick up where they left off as if nothing had happened to them. Apart from the fact that they are recovering from an injury, their absence will also have resulted in a decrease in their task fitness. It should be kept in mind that during the absence, they may have undergone surgery on a limb or may have had injections into the limb in an attempt to resolve the condition. Anything other than a gradual reintroduction to the work regime could result in a reappearance of the symptoms. In addition, a return to the same unchanged working conditions is likely to prompt a resurgence of symptoms if those conditions prompted the appearance of symptoms in the first instance.

It was stated at the outset of this chapter that often it is very simple aspects of the working environment that are the likely causes of ULDs. In most cases, these problematic features can be identified easily, and this can be done as part of a standard process of risk assessment. This system of risk assessment should identify potential problem areas before they actually become full-blown problems. Regardless of how proactive a company might be, it is unlikely that ULDs in the workplace will be eliminated altogether. However, it is possible to reduce them significantly through relatively simple measures.

11.4 Backache

The issue of backache has been dealt with several times in this book in the chapters relating to posture, workstation design, and manual handling. Office workers frequently complain about backache, usually in the lower back or lumbar region. Their pain can usually be connected with the adoption and maintenance of poor working postures and sitting for long periods without leaving their workstations. Back pain can also be connected with poor workstation design, particularly, poor chair design, that does not allow users to adjust their chairs so that they have fully supported backs, and poor use of the chair, which is usually a result of a lack of training and follow-up support by managers who should encourage proper use of the equipment. Finally, the manual handling of loads cannot be overlooked as a source of backache when office workers perform significant amounts of handling, often done without any appropriate training.

The only successful means of combatting back pain in the office is to provide users with suitable seating and train them in how to adopt appropriate postures when working. This needs to be done in conjunction with permitting computer users to leave their workstations at regular intervals so they can walk around. This does not have to be 'non-value-added' time. Companies can arrange the work so that when individuals leave their seats, they are walking to a shared printer, collecting files stored in a separate room, using the photocopier, collecting their own mail from a communal in-tray at a distance from their desks, and so on. In addition, if office workers are required to handle any loads, they need to be given the training to do so safely; this training needs to focus on the typical loads they handle during the course of their work.

11.5 Visual Fatigue

The term 'visual fatigue' is a far better descriptor of what is occurring when an individual is working than the term 'eyestrain'. The term 'eyestrain' suggests that the eyes have become damaged in some way, but the reality is that the eyes are more likely to be tired from the effort of trying to see. The symptoms of visual fatigue are typically short-lived and reversible, and reversing them is reliant on steps being taken to tackle provoking factors in the workplace.

Typical symptoms of eye fatigue include redness of the eyes, itching in the eyes, blurred vision, double vision, aching around the eyes, watering eyes, headaches, and, sometimes, nausea. The symptoms are considered to be caused by overuse of the muscles around the eyes, which are responsible for positioning and controlling the eye as people read the screen, look at the keyboard, and refer to documents. If an individual works for long periods without interruption, the muscles of the eye become fatigued like any other muscle in the body. The main causes of visual fatigue are poor screen positioning relative to overhead lights and windows, glare, inappropriate viewing distances, badly designed screen displays with characters that are difficult to read, unsuitable environmental conditions that result in

It is worthy of note that research has demonstrated that stress can increase muscle activity during display screen work (Lundberg et al. 2002), and it also tends to increase the forces applied to the computer mouse and leads to more rapid wrist movements in computer users (Wahlström et al. 2003). This has implications in terms of the development of ULDs.

11.7 Summary

- Although there is an association between computer work and conditions affecting the neck and upper limbs, the important factor is not that computer equipment is used, but rather how it is used.
- Limbs work most efficiently when moving within a comfortable range.
- Overextension of a joint or adoption of an irregular posture is likely to result in a ULD, if performed repeatedly or for a sustained period.
- ULDs can occur anywhere from the fingertip, along the arm, into the shoulder, and the back of the neck.
- Typical symptoms of a ULD include pain, aching, swelling, pins and needles, tingling, tenderness, feelings of heat, loss of grip strength, and clumsiness.
- The sooner ULD sufferers receive treatment, the more likely they are to make a successful recovery.
- Without intervention, a ULD sufferer may experience impairment of limb function and may be left with a residual disability.
- Tenosynovitis is a condition affecting the wrist area. Highly repetitive, forceful and/or irregular wrist positions are associated with its onset. Symptoms include aching, tenderness, swelling, reduced grip strength, and audible creaking on movement.
- Stenosing tenosynovitis is associated with overuse of the wrist and causes a clicking or pulling sensation on extension of the fingers or thumb.
- De Quervain's tenosynovitis is associated with prolonged or repeated effort, unaccustomed work, forceful gripping combined with deviation of the wrist and heavy reliance on the use of the thumb.
- Trigger finger is caused by overuse of the finger flexors and is associated with repeated or sustained gripping of an object with a hard or sharp edge.
- Carpal tunnel syndrome results from compression of the median nerve. Forceful gripping, irregular wrist postures, and vibration are

associated with its development. Typical symptoms include numbness and tingling in the hand and night waking.

- Dupuytren's contracture is congenital in some cases, but is also associated with repeated minor trauma or injury to the hand. It causes the ring and little fingers to bend toward the palm.
- Vibration white finger is associated with the use of vibrating equipment. It is caused by impairment of the blood circulation to the fingers, and symptoms include blanching of the fingers, numbness, tingling, and feelings of cold in the fingers.
- Ganglion has no generally accepted associations, but its widespread occurrence suggests unsuitable working conditions.
- Epicondylitis results from overuse of the muscles that bend the hand up or down at the wrist. Forearm rotation and forceful movements of the wrist and forearm are also likely contributors.
- Tendinitis is associated with forceful work and repetitive operations, as well as with working with raised arms. It causes aching around the shoulder and limitation of movement.
- A frozen shoulder can result from repeated sustained overhead work and results in a gradual onset of pain, which is more prominent at night, and restriction of movement.
- Thoracic outlet syndrome results in pain on the underside of the arm and numbness in the fingers. It is associated with working with the arms raised, working with the shoulders pulled back and down, and with repeatedly bearing loads on the shoulders.
- Cervical spondylosis affects the neck and spine, causing pain in the neck and shoulders. It is associated with age and tasks requiring heavy loads to be carried on the shoulders.
- Osteoarthritis causes stiffness, aching, and restriction of movement. It occurs most often in joints that are subject to long-term stress.
- Non-work-related factors, as well as work-related factors, can result in ULDs.
- The main causes of ULDs are repetition, awkward posture, static muscle work, force, and duration of activity.
- Other contributory factors in the development of ULDs include lack of rest breaks, lack of task variety, inconsistent work rate, lack of training and supervision, working in cold environments, and vibration.
- Psychological factors are considered to be associated with the development of ULDs.
- Clear guidance on what a sufferer can do on returning to work following an absence resulting from ULDs should be sought from the treating physician.

- On return to work after an absence, a sufferer should not be expected to pick up where they left off.
- Backache can be related to the adoption and maintenance of poor working postures, poor workstation design, lack of training, and manual handling.
- Backache can be avoided by using suitable workstation furniture, using it correctly, and leaving the workstation at regular intervals. Appropriate manual handing training should also be provided where required.
- Visual fatigue results from the effort of trying to see.
- Symptoms of visual fatigue include redness of the eyes, itching in the eyes, double vision, blurred vision, aching eyes, watering eyes, headaches, and nausea.
- The main causes of visual fatigue are poor screen position relative to overhead lights and windows, glare, inappropriate viewing distances, badly designed screen displays, unsuitable environmental conditions, and extended screen work.
- Stress results from mismatches between the demands placed on individuals and their ability to meet those demands.
- Prolonged stress can lead to anxiety, depression, heart disease, back pain, and gastrointestinal problems.
- One person experiencing stress may affect the morale of other members of the workforce.
- Stress is closely associated with role ambiguity, role conflict, excessive demands, a gap between demands and capabilities, lack of autonomy, isolation from the process of change, and social interaction within the workplace.

References

Arvidsson, I., Arvidsson, M., Axmon, A., Hansson, G.-Å., Johansson, C. R., and Skerfving, S. 2006. Musculoskeletal disorders among female and male air traffic controllers performing identical and demanding computer work. *Ergonomics* 49(11), 1052–1067.

Aydeniz, A., and Gursoy, S. 2008. Upper extremity musculoskeletal disorders among computer users. *Turkish Journal of Medical Sciences* 38, 235–238.

Bridger, R. S. 2017. *Introduction to Human Factors and Ergonomics*. 4th ed. Boca Raton, FL: CRC Press.

Cagnie, B., Danneels, L., Van Tiggelen, D., De Loose, V., and Cambier, D. 2007. Individual and work related risk factors for neck pain among office workers: A cross sectional study. *European Spine Journal* 16, 679–686.

Conlon, C. F., and Rempel, D. M. 2005. Upper extremity mononeuropathy among engineers. *Journal of Occupational and Environmental Medicine* 47, 1276–1284.

Corlett, E. N., and Bishop, R. P. 1976. A technique for assessing postural discomfort. *Ergonomics* 19, 175–182.

Dennerlein, J. T., and Johnson, P. W. 2006. Different computer tasks affect the exposure of the upper extremity to biomechanical risk factors. *Ergonomics* 49(1), 45–61.

Donoghue, M. F., O'Reilly, D. S., and Walsh, M. T. 2013. Wrist postures in the general population of computer users during a computer task. *Applied Ergonomics* 44, 42–47.

Gallagher, S., and Heberger, J. R. 2013. Examining the interaction of force and repetition on musculoskeletal disorder risk: A systematic literature review. *Human Factors: The Journal of the Human Factors and Ergonomics Society* 55, 108–124.

Gallagher, S., and Schall Jr, M. C. 2017. Musculoskeletal disorders as a fatigue failure process: Evidence, implications and research needs. *Ergonomics* 60(2), 255–269.

Gerr, F., Monteilh, C. P., and Marcus, M. 2006. Keyboard use and musculoskeletal outcomes among computer users. *Journal of Occupational Rehabilitation* 16, 265–277.

Gerr, F., Marcus, M., Ensor, C. et al. 2002. A prospective study of computer users: 1. Study design and incidence of musculoskeletal symptoms and disorders. *American Journal of Industrial Medicine* 41, 222–235.

Gerr, F., Marcus, M., and Monteilh, C. 2004. Epidemiology of musculoskeletal disorders among computer users: Lesson learned from the role of posture and keyboard use. *Journal of Electromyography & Kinesiology* 14, 25–31.

Griffiths, K. L., Mackey, M. G., and Adamson, B. 2007. The impact of a computerized work environment on professional occupational groups and behavioural and physiological risk factors for musculoskeletal symptoms: A literature review. *Journal of Occupational Rehabilitation* 17, 743–765.

Juul-Kristensen, B., Laursen, B., Pilegaard, M., and Jensen, B. R. 2004. Physical workload during use of speech recognition and traditional computer input devices. *Ergonomics* 47(2), 119–133.

Karlqvist, L., Wigeaus T. E., Hagberg, M., Hagman, M., and Toomingas, A. 2002. Self reported working conditions of VDU operators and associations with musculoskeletal symptoms: A cross-sectional study focusing on gender differences. *International Journal of Industrial Ergonomics* 30, 277–294.

Lee, D. L. McLoone, H., and Dennerlein, J. T. 2008. Observed finger behaviour during computer mouse use. *Applied Ergonomics* 39, 107–113.

Levanon, Y., Gefenc, A., Lermand, Y., Givona, U., and Ratzon, N. Z. 2012. Reducing musculoskeletal disorders among computer operators: Comparison between ergonomics. Interventions at the workplace. *Ergonomics* 55(12), 1571–1585.

Lundberg, U., Forsman, M., Zachau, G. et al. 2002. Effects of experimentally induced mental and physical stress on motor unit. *Work Stress* 16, 166–178.

Onyebeke, L. C., Young, J. G., Trudeau, M. B., and Dennerlein, J. T. 2014. Effects of forearm and palm supports on the upper extremity during computer mouse use. *Applied Ergonomics* 45, 564–570.

Putz-Anderson, V. 1992. *Cumulative Trauma Disorders: A Manual for Musculoskeletal Diseases of the Upper Limbs*. London, UK: Taylor & Francis Group.

Van Eerd, D., Hogg-Johnsona, S., Mazumdera, A., Colea, D., Wells, R., and Moore, A. 2009. Task exposures in an office environment: A comparison of methods. *Ergonomics* 52(10), 1248–1258.

Violante, F., Isolani, L., and Raffi, G. B. 2000. Case definition for upper limb disorders. In: Violante, F., Armstrong, T., Kilbom, A. (Eds.), *Occupational ergonomics: Work Related Musculoskeletal Disorders of the Upper Limb and Back*. London, UK: Taylor & Francis Group.

Wahlström, J., Lindegard, A., Ahlborg, Jr, G., Ekman, A., and Hagberg, M. 2003. Perceived muscular tension, emotional stress, psychological demands and physical load during VDU work. *International Archives of Occupational and Environmental Health* 76, 584–590.

Yang, J. F., and Cho, C. Y. 2012. Comparison of posture and muscle control pattern between male and female computer users with musculoskeletal symptoms. *Applied Ergonomics* 43, 785–791.

12

Disability

12.1 Introduction

The starting point for this chapter has to be a definition of what constitutes a disability. A disability is a physical or cognitive impairment that has a significant and ongoing negative impact on an individual's ability to perform standard everyday activities that would normally be carried out frequently. Physical impairments relate to problems with mobility, manual dexterity, coordination of movement, ability to lift and carry items, and sensory impairments that affect hearing and vision. Cognitive impairments relate to problems with memory, concentration, comprehension, and perception. Impairments can be congenital (present at birth) or can result from aging, injury or disease.

Disabilities take very diverse forms. Despite that, an employer should be able to provide a working environment that is as appropriate to the needs of the disabled person as it would be for someone without a disability. This is not necessarily an easy or straightforward process. It requires an appreciation of the limitations that a disability brings, as well as developing a knowledge of what specialised equipment or systems are available to reduce the impact of the disability.

This chapter will offer some advice on what disabled individuals need to enable them to do their work as safely and comfortably as possible. It will not provide a breakdown of every adjustment that should already be in hand to accommodate them, such as lower-level light switches, handrails, accessible washroom facilities, suitable parking arrangements, and so on. It simply focusses on assistive and adaptive technologies that enable a disabled person to work. Assistive technologies are products or equipment that can increase, maintain, or improve the functional abilities of a disabled person. Accessible design has been focussed on developing assistive technology to make products or systems easily usable by people with disabilities (Kim et al. 2016). Adaptive technology is any hardware or software created or modified to enable a disabled person to use a computer system. The use of these technologies is intended to increase disabled peoples' access to work.

To be able to respond appropriately to specific individual needs, an employer needs to understand the impact of each type of impairment on the work to be done. People with reduced strength, coordination, mobility, and control of their hands and fingers find it difficult to manipulate and manoeuvre objects, such as a mouse or telephone handset. If their range of motion and their upper body strength is limited, they will be unable to perform any task requiring force, such as lifting, and they will be limited in the extent to which they can reach forward or upward. If they have difficulty in controlling the movements of their limbs, particularly their arms, they will encounter problems in gripping, moving, and positioning objects, which has implications for mouse use. If head movement is difficult, this will limit the extent that a person can position and reposition the head relative to a display or document. A hearing-impaired person may have difficulties with verbal communication, which has implications for meetings, briefings, training, and so on. Visual impairment will obviously have an impact on the ability to decipher printed or displayed information in documents and screen displays. A cognitive impairment may result in an individual being unable to comprehend information presented to them in either a written or verbal format. A business might employ an individual who has a combination of impairments, which makes providing a suitable working environment a more challenging task. By establishing what impairments an employee has and how they make the performance of certain aspects of a task difficult, an employer will be able to find the solution that will enable the disabled person to work effectively. Dealing with disabilities differs from accommodating the non-disabled, who can usually be accommodated with an off-the-shelf approach; an employer has to seek specific solutions for each individual disabled person. Although disabled people will have undergone medical evaluations that specify their level of impairment, these evaluations do not, as a rule, stray into the area of what is required to enable people to continue working successfully.

There are some general guiding principles that employers should follow when sourcing solutions to impairments which affect how work can be done. Any solution they identify should increase the disabled person's independence within the workplace so that the work is completed more efficiently, and this, at the same time, should increase the employee's satisfaction. Enabling technology may allow some disabled people for the first time to retrieve information that was once inaccessible in books and magazines. Ensuring that the solution is in a usable format and easily learned will allow the disabled person to complete their work effectively. It should be compatible with any other devices in use and should be flexible enough to allow for variants of the work to be performed. The solution should be reliable on a day-to-day basis and should not present any foreseeable harm to the user. A degree of portability in the solution will ensure that users have more freedom over where and how they work, as opposed to being fixed in one location. Disabled people not only need to be physically comfortable when using the solution, they also need to feel emotionally comfortable using it in full view of their work colleagues. Employers should bear in mind that many of the choices employed for their disabled employees will also benefit the non-disabled workforce. For instance, shortcuts on the keyboard will reduce the amount of mouse and keyboard work performed by everyone.

12.2 Workstation Arrangements

Some individuals may come to work in wheelchairs. For many, these provide mobility and offer support to the upper body and legs. Most office workers are provided with fixed-height desks that tend to be around 720 mm in height, but wheelchair users may find that this does not offer sufficient clearance for them to get the chairs under the desk or to get knee clearance. As a consequence, they will have to sit at a distance from the leading edge of the desk and lean forward to reach the keyboard and mouse. Desks around 725 to 750 mm high are likely to offer sufficient clearance, if desk arms are used, although some believe that a desk height of 800–850 mm might be better for certain wheelchairs, particularly those fitted with joysticks. The latter higher levels will result in the keyboard being presented at a high level, which will be fatiguing for the upper limbs when the keyboard is in use. Trialling desks before purchasing them is the best way of determining whether a specific desk is suited to a particular individual and their wheelchair. Providing a height-adjustable desk would offer total flexibility over how the desk is set up and would cater to a wider range of disabilities. Electrically powered height adjustment should be considered for individuals who may find it difficult to grip an adjustment handle or who may not have sufficient strength to rotate one.

Some ambulant-disabled people may find that raising a chair to set themselves at a suitable height relative to a fixed-height desk is not a viable option, because they may have a condition that results in their legs not being reliable in terms of weight bearing. If their feet are not in contact with the floor, when they try to stand up by first sliding forward on the seat and placing their feet on the floor, they may find that their legs are unable to bear their weight, and they may fall. In this situation, a height-adjustable desk would allow them to lower the chair so that their feet are always firmly in contact with the floor; then they could lower the desk to an appropriate height relative to their sitting position.

The employee's wheelchair should have armrests that allow access to the desk, so fold-back or removable armrests are beneficial, unless the wheelchair has been fitted with specific desk arms. If the armrests are problematic, consideration should be given to using a keyboard tray. This should still allow for clearance of the joystick to the left or right of the tray. Alternatively, the user could be provided with a lap tray, which is a flat surface attached to a bean bag. The bean bag rests on the thighs and forms around them, allowing the tray to be positioned horizontally so that the keyboard can be supported on its surface.

A side-to-side knee-hole clearance 800 mm wide should be sufficient to allow the wheelchair to be located under the desk. This may necessitate the removal of drawer units. Ideally, there should be complete knee clearance from front to back under the desk, which may necessitate the repositioning or removal of modesty panels. At the very least, a minimum clearance of 600 mm is required from front to back under the desk. A minimum clearance of 700 mm between the undersurface of the desk and the floor is required.

The amount of space around a desk and any other areas utilised by people in wheelchairs may need to be increased to allow for a suitably sized turning circle. The chair will need sufficient space to be pushed back from the desk, then swivelled to the left or right before moving forward. An unobstructed space behind the desk of about 1000 mm is required. This should be increased to 1550 mm if two wheelchair users are working in the same area. If two wheelchair users are working back to back, and it is envisaged that a third wheelchair user may need to move between them when transferring from one work area to another, the space between desks should be 2050 mm. If a wheelchair user is working within an area where they have to access different shelves or cupboards, such as when sorting items in a storeroom, they should have an unobstructed floor space of 1500 × 1500 mm to turn around between facing shelves or between the wall and facing shelves. These dimensions should be applied in any other areas where an individual using a wheelchair might be expected to work or socialise.

If a wheelchair user is required to collect files and reference folders or any other material on a regular basis, efforts should be made to locate these items close to the employee's working location rather than in a remote storeroom or shelf. Repeated movement and repositioning of the wheelchair may be exhausting for some individuals. Shelves located above the desk should be set no higher than 1150 mm above the floor to allow a wheelchair user to access them with ease. If the user is considered ambulant-disabled, the shelves above the desk can be positioned 1620 mm above the floor. Anything that needs to be picked up or handled should be as close as possible to the individual to avoid reaching, which may be difficult or impossible for some.

Some wheelchair users may choose to transfer from their wheelchairs to task chairs when working at their desks. The task chair should be completely stable during this transfer. It should be kept in mind that many disabled users rely on the armrests for leverage and the chair should not be liable to tip up. Chairs with a standard five-star base may not provide enough stability, and alternatives should be considered. It is possible to purchase chairs with more than five feet on the base.

For many people, their disabilities (for instance, hearing or visual impairments) will not have an impact on the type of seat they use while working.

However, a number of other disabilities may necessitate the use of specifically selected chairs. The chair may need to offer a greater range of adjustment or more extensive support, such as a higher backrest or headrest. The available ranges and prices of these chairs are vast, and some specialist advice, in conjunction with the user's input, is advisable when putting together a short list of possible chairs to trial. It should not be assumed that a 'special' chair will be required in all cases. Often, a simple yet well-designed chair is all that is required. The important factor is that the individual responsible for identifying possible chairs needs to understand what the disabled user needs in terms of assistance and what constitutes good design in a chair. Once a chair has been selected, consideration has to be given to whether footrests are required; more sophisticated footrests may be needed that offer a larger surface area, have a greater range of adjustment, or can be altered through a pneumatic lift or electrically powered mechanism. The final choice should be driven by the user's needs and capabilities.

12.3 Computer Work

Wobbrock et al. (2011) have promoted what they refer to as 'ability-based design'. They feel that a user's ability to use a system should be considered when developing any systems of work, rather than focussing on his or her disabilities. Whatever product the individual uses to enable them to do their work should be 'aware' of their abilities and provide a suitable interface for these abilities before they start work. According to ability-based design, the standard system should adapt or be adapted to people with disabilities, rather than the disabled people adapting to use the standard system.

The standard interface with a computer is through a keyboard and input device, such as a mouse or tracker ball, that need quite precise manipulation. Many people with disabilities cannot use these devices because their range of movement is limited or they lack the strength required to depress the keys or click the buttons on a mouse; or they may not be able to hold the mouse steady while they try to locate the cursor and click the button because of uncontrolled hand movements or tremours. These individuals need to be able to access the system using the keyboard only or by using a specialised switch. Other users may have minimal movement capability and may rely on on-screen keyboards that can be activated by using an optical head pointing device or eye tracking device. Voice recognition software is another option.

Some users may be able to use only one hand to interact with their computer system. Using a standard keyboard will result in them having to cover a large area, and this will involve large movements of the hand and arm, which will be fatiguing. Using a compact keyboard, which is a slightly scaled-down version of a standard keyboard similar to a laptop keyboard, requires smaller movements of the hand and arm to reach individual keys. The dimensions are reduced by removing the number pad and by reducing the spacing around the function keys, rather than reducing the size of the alpha keys; this ensures that they are not more difficult to depress or are likely to be depressed inadvertently. The compact size of these keyboards also makes it more likely that they will fit between the arms of a standard wheelchair. They are also suited to the needs of individuals who depress the keys using a head pointer or mouth stick because their smaller dimensions reduce how far the head has to be moved. Keyboards used in this manner may need to be presented to the user at an elevated level to reduce the need to bend their heads forward; this can be achieved by using an articulating clamp system.

Keyboard users would normally be advised to position their hands above the keys and almost hover as they move their fingers between keys. Some disabled users are unable to hold their hands unsupported and may need to rest them on the keyboard. This may result in unintended activation of the keys. Some other disabled users may not be able to avoid pressing more than one key at a time. To avoid this, the keyboard can be fitted with a keyguard, an inflexible plate placed over the standard keyboard. The guard has holes that line up exactly with the keys. The user can lean on the guard without causing inadvertent depression of keys, and the holes assist the user to depress only single keys. Alternatively, larger keyboards can be purchased that already have a keyguard built in, in effect presenting the keys slightly below the surface of the keyboard, and these are suited to people who find it hard to use the standard-sized keyboard effectively. The larger area of these keyboards provides a bigger target for some disabled users to aim for, which increases the accuracy of their key depressions. Other keyboard types are available, such as flat keyboards with pressure-sensitive overlays that offer a range of keyboard layouts. Overlays are particularly useful if the user is only likely to need access to a limited number of keys or functions. As with any other piece of equipment, some research is required to identify the right equipment for the individual.

If a standard keyboard is used, the way it operates can be altered to make it more usable. For instance, the use of 'sticky' keys eliminates the need for several keys to be depressed simultaneously, which is helpful for users who can only use one finger to depress keys. Sticky keys help with the use of shift, control, and alt keys. For instance, if a user wants to type a word beginning with a capital letter, rather than holding down the shift key and depressing the letter to be capitalised, the sticky key (the shift key) is pressed first and remains depressed until the letter is depressed. Filter keys are a means of determining how long a key has to be depressed before the letter is presented on the screen or before it is repeated. This is helpful in situations where an individual might inadvertently lean on keys or may depress the key a second time unintentionally because of a tremour. Mouse keys allow the on-screen cursor to be repositioned using the keys of the number pad only. Users who can use only one hand, but can use each of their fingers independently, may benefit from trying a chord keyboard. This contains only a few keys and relies on the user being able to depress several keys simultaneously in musical chord fashion. Using predictive software can speed up typing rates for some keyboard users. They simply type in the first few letters of a word and are provided with a choice of words from which they select the target word. This is most effective if the user is likely to use longer words.

Virtual keyboards are widely used to help people with disabilities to interact both with desktop and laptop computers. Most of these keyboards have a set of features that facilitate the writing process, such as word completion methods (Li and Hirst 2005). A virtual keyboard for mobile devices, called BigKey has been developed (Faraj et al. 2009) to be used by people with motor disabilities. BigKey makes the keys of a virtual keyboard easier to access as it predicts what the next likely key to be used will be. There are alternatives to virtual keyboards that simplify text entry in mobile devices including interaction methods which work by recognising gestures (Wobbrock et al. 2004).

A number pad that is an integral part of the keyboard may not be useful for individuals who can use only one hand to operate the keyboard. Typically, number pads are presented to the right of the keyboard, which will cause a left-handed user to reach awkwardly across the body midline. Using separate number pads is a more flexible option.

Simply changing the settings on the mouse may make it easier to use. Ensuring that it operates slowly and steadily will make it easier to position and use. Setting the mouse so that it leaves a trail behind it on the screen as it moves will allow a visually impaired person to follow its movement more easily. Changing the functions of the mouse can also be helpful. For instance, a drag lock can be installed so that the button on the mouse can be clicked to lock onto an item on the screen and pull it into position without needing to hold down the button. Users simply click the button again when they want to drop the screen item into position.

Using a standard computer mouse may not be an option for some users. There are many alternatives available, many of which are discussed in Chapter 6, and some of these are designed specifically with the disabled user in mind. For instance, if users have difficulty with fine motor skills, such as executing small and controlled movements of the fingers, they can try a large trackball. These are designed to incorporate a large ball and two large buttons that can be operated using gross movements. These can also be operated by the feet if users cannot operate them with their hands.

If the user is unable to use either a mouse or keyboard, a switch can be an alternative. This is a button that can be activated by any controlled movement of the body, and that sends a signal to the computer. When activated, this can be used to drive a range of software packages. A user who has very limited movement can employ a system that allows control of the computer system through eye movements. These are referred to as eye tracking systems.

Gaze interaction enables the user to have hands-free control of computers. However, pointing to and selecting small targets using gaze alone is difficult because of the limited accuracy of gaze pointing (Skovsgaard et al. 2011).

For some individuals who have physical and cognitive impairments, using desktop touchscreens can be more effective than using keyboards and mice. However, it has to be kept in mind that some touch characteristics, i.e. how the individual touches the screen, may be affected by their type of disability (Irwin and Sesto 2012). For instance, the type of motor control disability may be important as differences in touch characteristics have been found between users with cerebral palsy and users with a tremour-related disability. It has been established that individuals with cerebral palsy have increased impulse and dwell time (dwell time relates to contact time on the button) when touching a screen (Irwin et al. 2011). Individuals with movement difficulties have less accuracy when aiming at specific areas of the screen and have a longer deceleration phase, i.e. it takes longer for them to slow their movement down once they move a limb in a particular direction. They also take longer to complete the task (Chen et al. 2013). Touchscreen applications that include the use of large icons that are easier to detect result in more successful use of a touchscreen. Disabled users have demonstrated improvement when they are given the opportunity to increase button size (Chen et al. 2013). It is also the case that a more intuitive flow of information can be helpful.

The disadvantages of using touchscreens are that they can be difficult for some people with specific physical disabilities to use, as well as for people with severe visual impairment, who may rely on receiving feedback through touch (such as on a keyboard), and, they may be difficult to use by people with poor literacy skills. To counter these problems, some touchscreens can be activated by using mouth or head sticks, and some can be altered to provide audio feedback for people with visual impairment. A system called NavTap (Guerreiro 2010) allows blind people to use mobile devices by inputting text using directional keys that provide audio feedback for each key pressed; the directional keys navigate through a predefined arrangement of the alphabet that is easy to remember. Another system called DigiTaps (Azenkot et al. 2013) has been developed in order to allow blind people to use touchscreen smartphones with minimal audio feedback. Users are able to enter a digit by tapping or swiping the screen with one to three fingers. They perform one or more gestures to achieve a desired digit, according to a specific code. This method provides two kinds of haptic feedback (i.e. feedback through the sense of touch), allowing users to know which gesture is registered based on vibrations of the device. This system is particularly useful if the user is dealing with sensitive material or is a little self-conscious of using a system with an audible form of feedback.

There are added issues to consider if the individual with a disability is attempting to use a mobile device, such as a smartphone. Individuals with motor disabilities in particular have trouble in interacting with small computer devices, because they usually do not have enough hand coordination to manipulate them effectively (Godinho et al. 2015). Almost all virtual keyboards available on mobile devices are too small to be used accurately by people with hand coordination problems. A system called EasyWrite, which has been designed with the intention of reducing the text entry problems encountered by motor-disabled persons when using mobile devices (Godinho et al. 2015), combines the concept of scanning systems with navigation functionalities in order to display bigger keys.

Scanning systems are a common means to help people with motor disabilities to interact with computers. This system presents a set of options to the user on the computer screen, whereupon a cursor moves through the options, one at time, at a pre-determined rate. When the cursor reaches the desired option, they press the button on an input device. Switches with one or two big buttons that are very easy to manipulate are usually used with this system.

Virtual keyboards can also incorporate scanning systems. They can either use sequential or group scanning. In the former, the cursor moves sequentially through all available options until the user presses a button to make a selection. Some users can find this process to be long and exhausting if they have to make a lot of selections. The use of group scanning can get round this problem by arranging the options into groups. Rather than navigating through individual options, the cursor moves through groups of options until the user chooses a group. At this point, the system then moves through a series of different tiers of subgroups until they find the specific individual option required.

Voice recognition software (VRS) has been discussed in Chapter 6. From the disabled person's point of view, it offers a hands-free means of interacting with a computer system. It is by no means an easy and intelligent exchange between the person and the computer; it is a rather complex system that works effectively if the right hardware is available, if users have been given the right kind of training, if they use a suitable technique, and if they are very patient. This system can even be used satisfactorily by vision-impaired people if a screen reader is used to send the displayed information to a voice synthesiser. This version of VRS is referred to as a voice-in-voice-out system. The difficulty with this system is that although, phonetically, a word may sound correct when read back to the listener, it might not be spelled correctly on the screen. For instance, listeners will not know when they hear the sentence 'Their files are over there?' whether the words 'their' and 'there' have been spelled correctly. In addition, vision-impaired users may become disoriented if they issue a command but are unexpectedly taken to a completely different area of the system than they had planned. Voice-out systems are often used in conjunction with notetakers, which consist of a portable Braille keyboard made up of six keys and a space bar, or a standard QWERTY keyboard, which the visually impaired person uses to make notes. These keyboards do not have screens attached. They are useful when users attend meetings and need to record the proceedings. The notetaker can be connected to a printer or a Braille embosser.

Before going down the road of using voice-in-voice-out systems, the visually impaired individual might be better to try learning to touch type using a standard keyboard. Visually impaired people can be helped during this process by being provided with specially designed stickers placed over the keys, which present the letter on the key in a large, capitalised, bold format that is easier to decipher. These are only to be considered aids to learning, as opposed to permanent features during their work. There are also stick-on bumps that can be used to identify specific keys that act as markers to guide the visually impaired person around the remainder of the keyboard. These can be provided in a variety of sizes and colours. Large-print keyboards with letters four times the size of standard keyboard print are also available. The keys can also be produced in a variety of colour combinations, allowing the individual to choose the combination that best suits their needs. Large keyboards are also available with keys four times larger than standard keyboard keys.

If visually impaired users have difficulty in using the mouse accurately to locate the on-screen cursor, they can use keyboard shortcuts, whereby a combination of keys are depressed simultaneously to effect a result.

If visually impaired individuals are required to read printed material as part of their work, such as files or reference folders, optical character recognition systems can be used to scan the printed text and store it electronically to be read back to them by a speech reader. Alternatively, text can be magnified on screen using the appropriate software. Magnifying software can increase the size of text, icons, and menus up to 32 times. Some software is also font-smoothing, to prevent distortion as the image is increased in size, and colour-converting to present images that are easier to decipher. Because only part of the whole image is viewable on the screen once magnified, a larger screen is normally used with this kind of system. The system usually presents on screen the area of interest that is identified by the location of the mouse at the time the document is magnified so the user does not normally have to reposition the image.

Some people with visual impairment can read a screen display as long as some key points are followed, bearing in mind that 'visual impairment' does not necessarily mean blind. These key points follow most of the general guidance outlined in Chapter 5 relating to screen displays, but with a few additions. For instance, auditory support should be an integral part of the system, for example, having key words from menus spoken as the on-screen cursor is moved over them. Having greater control over settings will allow visually impaired users to adjust the presentation to suit their particular needs. In some instances, they may adjust font type and size so that they can simply recognise word shapes as opposed to read words. The visually impaired will find it particularly helpful if windows do not overlap and the use of multiple windows is limited. Not only will this assist them to read the screen more easily, it will also assist the screen-reading software to 'find' the window position. Making the cursor icon and on-screen cursor more visible through an increase in size and a more distinct shape is also useful. Simplification of the language used may also make it easier for the visually impaired to read and understand the display.

Individuals who have difficulties with communication and language, or who have learning difficulties, may need to be presented with online information in a different format. This also applies to users whose first language is not that of the country in which they work. They may need the system and its accompanying vocabulary and displays to be simplified—for instance, by finding a way of eliminating elaborate menus. The material may need to be presented in a different way to make it easier to follow and use. Presenting material in a different way is referred to as 'repurposing'. What is particularly helpful is the magnification of words as they are read or spoken. This is also useful to people with visual impairment. The use of symbols or icons, instead of lengthy text, is beneficial, and accompanying these with screen tips that pop up along with a choice of auditory support is likely to increase comprehension, as are shortening sentences and removing unnecessary jargon. Yesilada et al. (2015) believe that developing accessible and usable websites and online information will benefit not only those with disabilities, but those without them as well.

12.4 General Environment

Consideration needs to be given to whether floor coverings are a source of risk to wheelchair users or the ambulant disabled. Some coverings may make it more difficult to manoeuvre a manually propelled wheelchair, while others may become extremely slippery if wet. An ongoing, active programme of monitoring floor conditions is required, because the floor finish will change over time, owing to wear and tear, the use of cleaning products, and the way floors are cleaned. For instance, if there is a spill in a communal area, such as a kitchen or foyer, it is common for the business to call cleaners to attend the area with a mop and bucket. The cleaners erect signs that warn that cleaning is in progress and the floor might be wet. They then take their mops to the spill, which might be the size of a mobile phone, and methodically mop from side to side, ultimately spreading a film of moisture over an area that may be hundreds of times bigger than the original spill. This increases the area that could potentially be slippery. A far more effective way of dealing with such a small spill and minimising the risk of slipping would be to blot the liquid up with absorbent paper.

If carpet is in place, the pile and type of underlay influence the ease with which a wheelchair can be manoeuvred. Deep-pile carpets can also create problems for people using walking frames or canes, or who have some difficulty when walking unaided.

High-gloss finishes on floors may not necessarily cause slipping problems, but some ambulant disabled may perceive these floors to be slippery, which may reduce their confidence when moving about. High-gloss finishes can also create problems in terms of glare, as can using paper with a glossy finish. Office workers who have impaired vision may find that they are confused by glare from work surfaces as they try to move around the building. Glare and reflections also have an impact on how easy it is for the hearing-impaired person to lip-read and to understand sign language. General advice about glare and how to control it is outlined in Chapter 9. The use of colours when decorating the office should be viewed in terms of how they can be applied to distinguish between different sections of an office, for instance, so that the visually impaired person can distinguish the door more easily from the wall when trying to move from one section of the building to another. Colour selection is most effective when it is done on the basis of colour intensity and luminance, rather than on the basis of extremes on the colour spectrum. Visually busy patterns on floors or walls should be avoided because they may give a visually impaired person a distorted perception of distance.

Choosing office materials that reflect sound and ultimately lead to a noisy environment make it difficult for a hearing-impaired person to hear what is being said and creates confusion for a visually impaired person owing to reflected sound. If, on the other hand, materials with an extremely high absorbency value are used, this will produce an environment that muffles sounds which can also be confusing.

12.5 Summary

- A disability is a physical or cognitive impairment that has a significant or ongoing impact on how an individual can live and work.
- Physical impairments relate to problems with mobility, manual dexterity, coordination of movement, and the ability to lift and carry items.
- Cognitive impairments relate to problems with memory, concentration, comprehension, and perception.
- An employer needs to understand the limitations each type of disability brings, as well as have knowledge about the specialised equipment or systems available.
- Assistive technologies are products or equipment that can increase, maintain, or improve the functional abilities of a disabled person.

- Adaptive technology is any hardware or software created or modified to enable a disabled person to use a computer system.
- Any solution used should increase disabled peoples' independence at work, allow them to complete their work more efficiently, increase their level of satisfaction, be in a usable format, and be easily learned. It should also be compatible with any other devices in use, be flexible enough to cope with variation in the work, be reliable day-to-day, not present foreseeable harm, and have a degree of portability.
- Desks should be set at a height to accommodate wheelchairs if appropriate.
- Desks should have sufficient undersurface clearance to accommodate the wheelchair.
- Adjustable-height desks offer greater flexibility.
- Keyboard trays can be considered if the wheelchair user cannot get close to the desk.
- The space around a desk and in any area used by a wheelchair user should provide sufficient space for an adequate turning circle.
- Task chairs need to be selected to suit each individual, as do footrests.
- A user's ability to use a system should be considered when developing any systems of work, rather than focussing on his or her disabilities.
- Certain disabilities may make the use of a mouse or keyboard difficult. Various input devices and different styles of keyboard are available as alternatives.
- Standard keyboards can be adapted through software to function differently (e.g. sticky keys, filter keys, and mouse keys).
- Changing mouse settings can make it easier to operate.
- Touchscreens can be effective for individuals with physical and cognitive impairments.
- Eye tracking systems allow a user who has very limited movement to employ a system that allows control of the computer system through eye movements.
- Scanning systems are a common means to help people with motor disabilities to interact with computers.
- Voice recognition software offers a hands-free means of interacting with a computer system. A screen-reader can be used to create a voice-in-voice-out system. Reliance on this system for correct spelling is not recommended.
- Learning to touch type can be a better option for some than using voice recognition software.
- Large print on keys or large keys can assist visually impaired people to type, as can magnification and changing the font style and size.

- Auditory support that states key words is effective.
- Simplifying systems, vocabulary, and displays may assist people experiencing difficulties with communication and language.
- Developing accessible and usable websites and online information will benefit not only those with disabilities, but those without them as well.
- Floor coverings and surfaces should not cause problems for wheelchair users nor the ambulant disabled.
- High-gloss finishes, glare, and busy patterns should be avoided on walls and floors.
- Colour choices should help to distinguish between sections of the office.
- Noisy environments cause confusion for hearing-impaired and visually impaired people.

References

Azenkot, S., Bennett, C. L., and Ladner, R. E. 2013. DigiTaps: Eyes-Free number entry on touchscreens with minimal audio feedback. In: *Proceedings of the 26th Annual ACM Symposium on User Interface Software and Technology (UIST'13)*, pp. 85–90. New York: ACM.

Chen, K. B., Savage, A. B., Chourasia, A. O., Wiegmann, D. A., and Sesto, M. E. 2013. Touch screen performance by individuals with and without motor control disabilities. *Applied Ergonomics* 44, 297–302.

Faraj, K. A., Mojahid, M., and Vigouroux, N. 2009. BigKey: A virtual keyboard for mobile devices. In: *Proceedings of the 13th International Conference on Human-computer Interaction. Part III: Ubiquitous and Intelligent Interaction*, Vol. 5612, Lecture Notes in Computer Science, pp. 3–10. Berlin, Germany: Springer.

Godinho, R., Condado, P. A., Zacarias, M., and Fernando, G. L. 2015. Improving accessibility of mobile devices with easy write. *Behaviour & Information Technology* 34(2), 135–150.

Guerreiro, T. 2010. Assessing mobile-wise individual differences in the blind. In: *Proceedings of the 12th International Conference on Human Computer Interaction with Mobile Devices and Services (MobileHCI'10)*, pp. 485–486. New York: ACM.

Irwin, C. B., and Sesto M. E. 2012. Performance and touch characteristics of disabled and non-disabled participants during a reciprocal tapping task using touch screen technology. *Applied Ergonomics* 43, 1038–1043.

Irwin, C., Yen, T. Y., Meyer, R. H., Vanderheiden, G. C., Kelso, D. P., and Sesto, M. E. 2011. Use of force plate instrumentation to assess kinetic variables during touch screen use. *Universal Access in the Information Society* 10(4), 453–460.

Kim, H. K., Han, S. H., Park, J., and Park, J. 2016. The interaction experiences of visually impaired people with assistive technology: A case study of smartphones. *International Journal of Industrial Ergonomics* 55, 22–33.

Li, J., and Hirst, G. 2005. Semantic knowledge in word completion. In: *Proceedings of the 7th International ACM SIGACCESS Conference on Computers and Accessibility (ASSETS'05)*, pp. 121–128. New York: ACM Press.

Skovsgaard, H., Mateo, J. C., and Hansen, J. P. 2011. Evaluating gaze-based interface tools to facilitate point-and-select tasks with small targets. *Behaviour & Information Technology* 30(6), 821–831.

Wobbrock, J. O., Kane, S. K., Gajos, K. Z., Harada, S., and Froehlich, J. 2011. Ability-based design: Concept, principles and examples. *ACM Transactions on Accessible Computing* 3(3), 1–27.

Wobbrock, J. O., Myers, B. A., Aung, H. H., and LoPresti, E. F. 2004. Text entry from power wheelchairs: Edgewrite for joysticks and touchpads. In: *Proceedings of the 6th International ACM SIGACCESS Conference on Computers and Accessibility (ASSETS'04)*, pp. 110–117. New York: ACM Press.

Yesilada, Y., Brajnik, G., Vigo, M., and Harper, S. 2015. Exploring perceptions of web accessibility: A survey approach. *Behaviour & Information Technology* 34(2), 119–134.

13

Risk Assessment

13.1 Introduction

The risk assessment process is part of a whole health and safety management system. Assessments are a means for an organisation to identify where it is in terms of health and safety matters and they have a significant influence on the development of health and safety strategies. A distinction should be made between safety risks and health risks. Safety risks can lead to immediate injury, such as coming in contact with a corrosive substance, whereas health risks may not be apparent for a long time and may result from daily exposure to unsuitable working conditions over months or years, as can occur in the development of upper limb disorders. The nature of this book is such that this chapter will focus on health risks.

The purpose of a risk assessment is first to identify hazards in the working environment. A hazard is something with the potential to cause harm. Once a hazard is identified, the level of risk associated with that hazard is assessed. In other words, the assessor attempts to determine how likely it is that an individual will come to harm when exposed to those particular hazardous circumstances. The risk assessment process also allows risk control measures to be identified if already in place, or recommended if not in place.

When assessors are trying to identify hazards in the workplace, some are immediately obvious or are already well recognised so that their identification is straightforward, and there is probably a tried and tested method for reducing risk. However, this is not always the case, and assessors have to learn about what is likely to constitute a hazard in a given working environment. They can do this by attending specific risk assessment training courses, reading standard operating procedures relating to the tasks, reading available guidance and advisory information, reading trade association information, networking with others in similar environments, reviewing accident and absence statistics for the business, and asking outside consultants to assist. They will probably identify the risk through observation and by taking measurements, such as measuring the effort involved in pushing a trolley.

Having identified the hazards, the assessor has to assess the risk. This requires an understanding of the task and the working practices in the area being assessed. The assessor should determine whether the hazardous activity or situation presents a high, medium, or low level of risk to the workforce. This will help the assessor to prioritise how to tackle the programme of intervention. Obviously, higher-risk situations need to be tackled before low-risk situations. Some assessors may be uncomfortable about making a judgment regarding the level of risk of an activity or situation. These individuals may feel more comfortable using a risk matrix to identify the level of risk (see Table 13.1). This is a method of ranking hazards and risks. To use this matrix, the assessor decides what the outcome or consequence would be should a worker come in contact with the hazard. They decide whether the individual would encounter a minor injury, a moderate injury, or a serious injury in that case. Each of these three consequences is assigned a value: 1, 2, or 3. The assessor then decides how likely it is that a worker will actually come to harm, given the existence of the hazardous situation. As can be seen from Table 13.1, they can select either very unlikely, unlikely, or likely, and, again, these are assigned a value of 1, 2, or 3.

Risk can be defined as the severity of harm combined with the likelihood of occurrence. When the assessor cross-multiplies the preselected values in Table 13.1, the computation results in the total score falling into one of the shaded boxes in the matrix. The number values in these boxes, as well as the shading used, inform the assessor that the task is either low (1–2), medium (3–4), or high risk (6–9). Many variations of the risk matrix are used; some of these have up to ten categories to choose from in terms of likelihood rating and consequence rating. However, they end up with the same results. Quite often, keeping risk assessment on a simpler footing is more effective.

Having determined that the workforce is at risk, the assessor has to develop risk reduction strategies. The first consideration should be whether the hazard can be eliminated altogether. For instance, if the assessor

TABLE 13.1

Risk Matrix Used to Identify the Level of Risk

LIKELIHOOD				
Very Likely	3	3	6	9
Likely	2	2	4	6
Unlikely	1	1	2	3
		1	2	3
		Minor Injury	Moderate Injury	Serious Injury
		CONSEQUENCE		

identifies that lifting bottles of water weighing 18 kg into a water dispenser is high-risk, this task could be eliminated by providing a vending machine fed directly from the main's supply. If the hazard cannot be eliminated, it must be controlled. In the first instance, the assessor needs to consider using a lower-risk option, such as providing more suitable seating with a larger range of adjustment. Alternatively, they could consider protecting the workforce from the hazard by restricting their exposure to it. For instance, the assessor may suggest that more regular rest breaks are required.

It is always helpful if the assessor records on the assessment form what control measures are already in place. This will ensure that those responsible for introducing risk reduction strategies are aware of what is already in place and how effective the strategies are at reducing risk. Assessors should also record the time scales they feel are appropriate to reducing the risks they have identified. This will allow an appropriate timetable to be drawn up. They may also point to the individuals who need to be involved in the risk reduction process; for instance, they may suggest that the training department needs to be involved because workers have not received training in the use of their equipment.

Of course, the assessment is only part of a much larger system of controlling risk in the workplace. The whole organisation needs to be aware of and involved in the control of risks in the workplace. To that end, the organisation needs a specific policy on how they intend to tackle not only the risk assessment process but also the control of risk. This will require cooperation among people at many levels within an organisation. For instance, if the assessor decides that some people require new chairs, the individuals responsible for allowing expenditure should understand why this is necessary and permit it. Everyone who has some influence on the risk assessment and risk control processes should have a clear understanding of their role and responsibilities. Each of these individuals must have a level of competence compatible with what they are required to do. All of them should communicate clearly with their colleagues about what steps they are taking and how they will affect the areas of concern.

One of the major failings of many risk assessment processes is that the system comes to a halt once the assessor has completed the checklist. Completion of the assessment checklist is not an end in itself; it is just a part of the risk identification and reduction process. There has to be a drive to put measures in place to reduce risk. In addition, someone needs to ensure, having introduced a number of changes, that the outcomes are satisfactory. If not, the assessor needs to reconsider the recommendations.

Once the assessment process is complete, a review should take place at intervals to ensure that it remains valid. It is unlikely that an office will remain unchanged for very long. New people are recruited, new workstations are purchased or old workstations repositioned, new tasks are developed, and changes occur in the environment. All of this should prompt a review of the assessment.

13.2 Sample Checklists

There are many ways to assess risk. For instance, employers can use self-assessment, simple observational checklists, or more detailed analysis using video footage. Direct measurements of movements, forces, and postures can be taken using sensors. However, methods that require specialised equipment are costly and require extensive technical proficiency (David 2005), so not many employers will choose this option. Conversely, online self-assessment questionnaires are inexpensive but their consistency is problematic (Chiasson et al. 2015). Skilled independent specialist assessors are a medium-cost way to effectively assess risk (Laestadius et al. 2009; Pillastrini et al. 2007, 2010; Robertson et al. 2009).

An employer does not need to invest heavily, from a financial perspective, when trying to carry out assessments. They just need to ensure that the individual who does the assessment has been trained to carry out that form of assessment and knows what to look for, or look at, during that process.

The remainder of this chapter contains a series of sample checklists that can be used during the assessment of various aspects of the working environment. They are not intended to cover every element of a workplace that an organisation should be assessing; they relate to those areas that have been discussed generally in this book.

Table 13.2 is a checklist intended to assess the workplace for hazards that are likely to contribute to the development of musculoskeletal disorders generally. This includes upper limb disorders, backache, leg discomfort, and so on. This particular checklist should only be completed by an individual who has a thorough knowledge and understanding of what is likely to contribute to the development of such conditions. The reason for this is that the checklist does not ask a series of prompt-like questions. It is simply a record sheet to allow knowledgeable assessors to detail what they identify as problematic. The risk matrix described above should be used in conjunction with this record sheet.

Table 13.3 is a specific upper limb disorder (ULD) checklist. It is intended to assist the assessor in identifying whether there are any aspects of the work or working environment that have the potential to contribute to the development of ULDs. Table 13.4 outlines a checklist intended to be used for a display screen equipment-based task; it assists the assessor to identify whether any element of the workstation, work, or immediate work area is unsuitable for the individual. Should a business decide to use an online self-assessment form of risk assessment, they must ensure that the user has suitable training prior to completing the checklist. Employers should also include individuals who usually work away from the main office building in their assessments. Table 13.5 outlines a checklist relating to mobile devices so that the employer is aware of the exact circumstances in which it is used. Table 13.6 includes a manual handling checklist applicable to any manual handling task performed within the organisation's building, as well as away from the building, such as if removing loads from a vehicle. Table 13.7 is a specific and detailed pushing and pulling checklist. Finally, Table 13.8 contains a 'tick list' of general elements to be considered during any assessment; this is a useful tool to use during a preliminary walkthrough of an environment when a general overview is being collated.

TABLE 13.2

A Checklist Used for Identifying Hazards Likely to Contribute to the Development of Musculoskeletal Disorders

MUSCULOSKELETAL DISORDERS RISK ASSESSMENT RECORD SHEET

This risk assessment was undertaken by: Date:

Task/s covered by this assessment: Reference:

The potential harm from the assessed task is:

Upper Limb Disorder ☐ Back Injury ☐ Other: ....

The factors that contribute to the potential harm are:

-
-
-

The people who have the potential to be harmed are:

-
-
-

The **Likelihood** score for them being harmed in this way has been assessed on a scale of 1-3 to be: ☐

The **Consequence** score for them being harmed in this way has been assessed on a scale of 1-3 to be: ☐

Overall assessment of the **RISK** (Likelihood x Consequence) = ☐

Suggested remedies to eliminate or reduce the assessed risk:	Timescale in weeks: →
• • •	

Assessor's estimate of **residual risk** following implementation of suggested remedies: L..... x C..... = RISK ☐

Which Manager is responsible for the task or area covered by this assessment?

I **accept** this assessment for action as the Manager with responsibility for the task or area assessed.

Name: Signature: Date:

I am **'signing-off'** this assessment as the responsible Manager for the task or area covered by this assessment and confirm that steps have been taken to control the significant risks identified.

Name: Signature: Date:

TABLE 13.3
Upper Limb Disorder Checklist

UPPER LIMB DISORDER CHECKLIST			
Assessor:		Date:	
Employee:		Job Title:	
Department:		Location of business:	
Working hours:		Working days:	

Control measures already in place:

SUMMARY OF RESULTS

Is this employee considered to be at risk?	Yes		No	

Remedial action required to reduce risk:	Actioned on:	Actioned by:	Outcome:	Review date:

(Continued)

TABLE 13.3 (*Continued*)

Upper Limb Disorder Checklist

Indicate whether any element is present by ticking the appropriate box and supply any additional comments relevant to the assessment. If you answer YES to any questions you should suggest remedial actions. Remedial actions should be entered on the previous page along with any control measures already in place.

POSTURE - Does the operator:	YES	NO	COMMENTS
◆ Have their arm(s) outstretched in front of them repetitively or for prolonged periods?			
◆ Hold their elbow(s) away from the side of the body repetitively or for prolonged periods?			
◆ Hold their forearm(s) such that the hand is higher than the elbow with less than a 90 degree angle at the elbow?			
◆ Repetitively work with their wrists bent to the side for prolonged periods?			
◆ Repetitively move their wrist from side-to-side when working?			
◆ Repetitively move their wrist up and/or down?			
◆ Rotate the wrist to turn the palm of the hand upward or downward?			
◆ Use irregular or unnatural hand and finger postures?			
◆ Forcefully grip for prolonged periods or repetitively?			
◆ Lean their head forward when working?			
◆ Rotate their head to one side or from side-to-side repetitively or for prolonged periods?			
TASK ACTIVITY - Does the operator:			
◆ Use force when completing the task either repetitively or for prolonged periods?			
◆ Use force with the limbs at or near the extreme ranges of movement?			
◆ Use static effort to complete the task?			
◆ Carry out the same movements or sequence of movements repeatedly?			
◆ Carry out these movements rapidly?			
◆ Work for long periods without a break in activity?			
◆ Work at inconsistent rates on occasion?			
◆ Use abrupt movements when working?			

(*Continued*)

TABLE 13.3 (*Continued*)

Upper Limb Disorder Checklist

EQUIPMENT CHARACTERISTICS:			
◆ Do the design features of the mouse or other input devices cause the worker to adopt an awkward wrist or arm posture?			
◆ Do the design features of the desk prevent the worker from adopting a suitable posture?			
◆ Do the design features of the chair prevent the worker from adopting a suitable posture?			
◆ Does the screen display cause the worker to lean forward when reading it?			
◆ Does the worker lean on any sharp edges when working?			
ENVIRONMENTAL CONDITIONS:			
◆ Do the lighting levels make it difficult for the worker to see properly when working?			
◆ Do the lighting levels result in the worker having to alter their posture to 'avoid' shadows, glare or reflections?			
◆ Are workers ever exposed to extremes in temperature?			
◆ Are workers ever exposed to changes in air quality which may affect their performance or general feelings of well-being?			
◆ Are workers exposed to noise levels which may be stressful?			
PERSONAL FACTORS - Is the individual:			
◆ Complaining about pain, discomfort, pins and needles, tingling or swelling?			
◆ Wearing any form of support bandage on their upper limbs?			
◆ Working faster than is necessary?			
◆ Working through their breaks?			
◆ Working regular overtime?			
◆ Working to meet deadlines regularly?			
◆ Using jerky, un-smooth methods of work?			
◆ In need of further training to improve their skill and knowledge base and to discourage poor work practices?			
◆ Anxious, distressed, or dissatisfied?			
◆ Experiencing difficulties because of their gender?			

(*Continued*)

TABLE 13.3 (*Continued*)
Upper Limb Disorder Checklist

◆ Experiencing difficulties because of their stature?			
◆ Experiencing difficulties because of their level of ability?			
◆ Experiencing difficulties because of previous/current injury or ill-health?			
◆ Involved in the lifting or moving of a load? If YES, complete a separate Manual Handling Questionnaire.			

TABLE 13.4

Checklist to Assess a Desk-Based Computer Task

DISPLAY SCREEN EQUIPMENT (DSE) ASSESSMENT CHICKLIST			
Assessor:		Date:	
Employee:		Job Title:	
Department:		Location of business:	
Working hours:		Working days:	
Number of hours DSE used per day:		Restbreak allowance:	

Specific location(s) where DSE work carried out:	Brief description of DSE work in each location:

Control measures already in place:

SUMMARY OF RESULTS

Remedial action required to reduce risk:	Actioned on:	Actioned by:	Outcome:	Review date:

(Continued)

TABLE 13.4 (*Continued*)

Checklist to Assess a Desk-Based Computer Task

Tick either the YES or NO column and provide additional comments if needed or helpful. If you make any adjustments whilst completing the assessment record this in the "action taken by assessor" column.

DESK	YES	NO	Comments:	Action taken by assessor:
Is there sufficient space on the desk surface to arrange equipment and materials appropriately?				
Can all equipment and documents be reached easily by the seated user?				
Is there sufficient space at and under the desk to allow the user to move freely?				
Is the user likely to come in contact with any trailing cables?				
Is the desk in good condition?				
Is the desk (and surrounding surfaces) free of glare and reflections?				

(*Continued*)

TABLE 13.4 (*Continued*)

Checklist to Assess a Desk-Based Computer Task

CHAIR	YES	NO	Comments:	Action taken by assessor:
Does the chair have a five-star base with castors or glides?				
Does the chair swivel?				
Does the chair adjust for height?				
Has the user adjusted their chair height correctly relative to the desk and keyboard?				
Does the backrest adjust for height and tilt?				
Do all of the controls and adjusters work?				
Have seat angle and seat length been correctly adjusted (if features available)?				
Does the user know how to adjust the seat?				
Can the user sit in a comfortable position even if armrests are present on the chair?				
Has the user adopted a suitable posture?				
Do the user's feet touch the floor/footrest?				

(*Continued*)

TABLE 13.4 (*Continued*)
Checklist to Assess a Desk-Based Computer Task

DESKTOP SCREEN	YES	NO	Comments:	Action taken by assessor:
How many screens in use simultaneously?				
Does each screen swivel and tilt?				
Are the screens free from reflections/glare?				
Is the image stable?				
Are the screens cleaned regularly?				
Can brightness and contrast be adjusted?				
Is the user aware of how to use the brightness and contrast controls?				
Are screens used for equal periods of time?				
Are the screens, or is the priority screen, directly in front of the user when in use?				
Are the screens at a suitable height and distance from the user?				
Are the characters on the display easy to read from the normal seated position?				

(*Continued*)

TABLE 13.4 (*Continued*)

Checklist to Assess a Desk-Based Computer Task

If a laptop is used is it attached to a docking station or a separate keyboard and mouse? Is a document holder used in conjunction with the screen(s)? Is the user considered to be working appropriately if they are not using a document holder?				

MOBILE DEVICES	**YES**	**NO**	**Comments:**	**Action taken by assessor:**
Does the user work with mobile devices, such as tablets, as a normal part of their daily work? If so, complete a separate checklist. Do they use them in conjunction with any peripheral devices? Are users able to adopt a comfortable posture when using the device(s)?				

(*Continued*)

TABLE 13.4 (*Continued*)

Checklist to Assess a Desk-Based Computer Task

INPUT DEVICES	YES	NO	Comments:	Action taken by assessor:
How many keyboards in use?				
Is the keyboard separate from the screen?				
Does the keyboard have a tilt facility?				
Is the keyboard in good working order?				
Is the keyboard in an appropriate position relative to the user?				
Is there sufficient room in front of the keyboard for the user to rest their wrists?				
Is the keyboard operated without force?				
Is a separate numberpad in use?				
How many mice in use?				
Is the mouse kept close to the user?				
Can the user adjust the mouse settings?				
Does the user rest their hand regularly?				
Is a mouse mat necessary?				

(*Continued*)

TABLE 13.4 (*Continued*)

Checklist to Assess a Desk-Based Computer Task

SOFTWARE	YES	NO	Comments:	Action taken by assessor:
Can the user work at their own pace? Is the software easy to use? Has the user received suitable training?				

ENVIRONMENT	YES	NO	Comments:	Action taken by assessor:
Is the lighting appropriate for the work? Is the overhead light in a suitable position? Is the screen in a suitable position relative to the window? Do the blinds or curtains work effectively? Is the heating and ventilation adequate? Is the level of humidity adequate? Are noise levels acceptable? If working away from the office, can the user take steps to control the environment?				

(Continued)

TABLE 13.4 (*Continued*)

Checklist to Assess a Desk-Based Computer Task

ORGANISATIONAL FACTORS	YES	NO	Comments:	Action taken by assessor:
Has the user received instructions in the use and arrangement of their workstation furniture?				
Has the user received health and safety training in relation to their work and workplace?				
Is the user aware of who to approach if they are experiencing any difficulties relating to health and safety issues?				
Does the user have regular changes in activity which break up their DSE work?				
Does the user have regular restbreaks which break up their DSE work?				

NOTES: (Use this section to discuss any user health problems as well as for additional information).

This assessment has been completed in the presence of the user who has agreed the contents, unless otherwise stated in the notes above.

TABLE 13.5

Checklist to Assess Use of Mobile Devices

MOBILE DEVICE ASSESSMENT CHECKLIST			
Assessor:		**Date:**	
Employee:		**Job Title:**	
Department:		**Location of business:**	
Working hours:		**Working days:**	

List mobile devices in use:	**Specify daily usage in hours:**	**Specify weekly usage in days:**	**List type of work done with devices:**

Risk reduction measures already in place:

Further risk reduction measures required:	**Implemented on:**	**Outcome:**

(Continued)

TABLE 13.5 (*Continued*)

Checklist to Assess Use of Mobile Devices

Indicate the response to the question by ticking the appropriate box and supply any additional comments relevant to the assessment. Remedial actions should be entered on the previous page along with any control measures already in place.

USE - Does the employee:	**YES**	**NO**	**COMMENTS**
◆ Work with mobile devices as a normal part of their daily work?			
◆ Work with mobile devices regularly even if not required? Specify.			
◆ Work with a mobile device on their lap? Specify.			
◆ Work with hardcopy documents occasionally so as to avoid using mobile devices?			
◆ Have access to standard desk-top equipment to use instead of mobile devices? Specify.			
◆ Work with peripheral items, such as a separate keyboard, to make the use of the mobile device easier? List these items.			
◆ Work with mobile devices that have been selected with a consideration of their weight and size? Detail weight/size.			
◆ Have access to suitable rucksacks/bags/trolleys to transport their mobile device?			
◆ Find the mobile device heavy when travelling?			
◆ Limit what they carry in addition to the mobile device?			
◆ Feel that the peripheral items carried with the mobile device make the complete load heavy?			
◆ Regularly use a rucksack, wheelie bag or other means to transport the mobile device?			
ENVIRONMENT - Does the employee:			
◆ Work in forms of transport? If yes, what?			
◆ Have access to a table when travelling?			
◆ Have the ability to avoid using mobile devices while on transport?			
◆ Have the ability to access work zones in waiting areas prior to travel?			
◆ Work at home?			
◆ Work in other offices connected with the business? Specify.			
◆ Have the ability to contact offices connected with the business and request suitable working arrangements in advance of arriving?			
◆ Work in hotels and/or at conferences? Specify.			

(*Continued*)

TABLE 13.5 (*Continued*)

Checklist to Assess Use of Mobile Devices

TRAINING – has the employee:			
◆ Been trained on how to arrange the environment in which they find themselves so that they can work comfortably?			
◆ Been provided with specific advice relating to working within different forms of transport?			
◆ Been advised about the posture they should aim to achieve?			
◆ Been advised about all of the peripheral equipment they can use with the mobile devices and why they should be used?			
◆ Been advised where to order peripheral equipment?			
◆ Been provided with advice on how to position/hold mobile devices to reduce discomfort?			
◆ Been trained on how to adjust the screen brightness on their mobile device so they can adjust it to suit their surroundings?			
◆ Been advised to use standard desktop equipment when they have the opportunity?			
◆ Been advised about the frequency of break taking when using mobile devices?			
◆ Been advised about making written notes and voice recordings to reduce the amount of manual inputting done on the mobile device?			
◆ Been provided with manual handling advice and training relating to the movement of mobile devices?			
◆ Been advised about the appropriate rucksacks/bags/trolleys available to transport mobile devices?			
◆ Been advised about the most appropriate way to wear rucksacks/bags?			
◆ Been advised of the symptoms that might indicate they are developing a musculoskeletal disorder?			

(*Continued*)

TABLE 13.5 (*Continued*)

Checklist to Assess Use of Mobile Devices

ORGANISATIONAL FACTORS – does the employee:			
♦ Work for concentrated periods of time on the mobile device?			
♦ Work to tight deadlines? Specify.			
♦ Take regular breaks away from using mobile devices?			
♦ Have software enabled on their mobile device to alert them to the need for a break?			
♦ Provide information that confirms they are working according to agreed business protocols? Specify.			
♦ Know who to contact if they are experiencing problems such as pain or discomfort?			
PERSONAL FACTORS - has the employee:			
♦ Reported symptoms such as tingling, pins and needles, pain, aching, swelling in their hands or arms?			
♦ Reported discomfort in their back or neck?			
♦ Reported lower limb discomfort?			
♦ Complained of headaches or tired eyes?			
♦ Complained of general dissatisfaction with their working conditions?			
♦ Encountered any difficulties relating to their stature?			

TABLE 13.6
Checklist to Assess Manual Handling Operations

MANUAL HANDLING OF LOADS ASSESSMENT CHECKLIST			
Assessor:		**Date:**	
Employee:		**Job Title:**	
Department:		**Location of business:**	
Working hours:		**Working days:**	

Description of manual handling task(s) carried out:	Loads weights handled:	Frequency of lifts:	Carrying distance:

Control measures already in place:

SUMMARY OF RESULTS

Is this employee considered to be at risk?	Yes		No	

Remedial action required to reduce risk:	Actioned on:	Actioned by:	Outcome:	Review date:

(Continued)

TABLE 13.6 (*Continued*)
Checklist to Assess Manual Handling Operations

Indicate the level of risk present by ticking the appropriate box and supply additional comments relevant to the assessment. Any suggestions for possible remedial actions should be entered on the previous page along with any control measures already in place.

Questions to consider:	Level of risk			**Comments:**
	Low	**Med**	**High**	
Do **the tasks** involve:				
♦ holding loads away from trunk?				
♦ twisting?				
♦ stooping?				
♦ reaching upwards?				
♦ large vertical movement?				
♦ long carrying distances?				
♦ strenuous pushing or pulling? (complete separate checklist)				
♦ unpredictable movement of loads?				
♦ repetitive handling?				
♦ insufficient rest or recovery?				
♦ a work rate imposed by a process?				
Are **the loads**:				
♦ heavy?				
♦ bulky/unwieldy?				
♦ difficult to grasp?				
♦ unstable/unpredictable?				
♦ intrinsically harmful (e.g. sharp/hot)?				

(*Continued*)

TABLE 13.6 (*Continued*)
Checklist to Assess Manual Handling Operations

Consider the **working environment** – are there:				
◆ constraints on posture?				
◆ poor floors?				
◆ variations in levels?				
◆ hot/cold/humid conditions?				
◆ strong air movements?				
◆ poor lighting conditions				
Consider **individual capability** – does the job:				
◆ require unusual capability?				
◆ pose a risk to those with a health problem or a physical or learning difficulty?				
◆ pose a risk to those who are pregnant?				
◆ call for special information/training?				

	YES	NO	
Protective clothing:			
◆ Is clothing or personal protective equipment a hindrance?			
◆ Is additional personal protective equipment required?			

TABLE 13.7

Checklist to Assess Pushing and Pulling Operations

PUSHING AND PULLING OF LOADS ASSESSMENT CHECKLIST			
Assessor:		Date:	
Employee:		Job Title:	
Department:		Location of business:	
Working hours:		Working days:	

Description of pushing and pulling task(s) carried out:	Forces involved:	Frequency of operation:	Distance involved:

Control measures already in place:

SUMMARY OF RESULTS

Is this employee considered to be at risk?	Yes		No	

Remedial action required to reduce risk:	Actioned on:	Actioned by:	Outcome:	Review date:

(Continued)

TABLE 13.7 (*Continued*)
Checklist to Assess Pushing and Pulling Operations

Indicate the level of risk present by ticking the appropriate box and supply additional comments relevant to the assessment. Any suggestions for possible remedial actions should be entered on the previous page along with any control measures already in place.

Questions to consider:	Level of risk			Comments:
	Low	Med	High	
Do **the tasks** involve:				
◆ high forces to start the movement of the load (measure forces unless obviously trivial)				
◆ high forces to maintain movement of the load (measure forces unless obviously trivial)				
◆ jerky movement when controlling the load?				
◆ moving of the load into a precise location or around obstacles?				
◆ one–handed pushing/pulling?				
◆ using the hands below the waist or above shoulder height?				
◆ pushing/pulling the load quickly?				
◆ transferring the load across a distance of more than 20m				
◆ pushing/pulling repetitively or for prolonged periods?				
The **load or object** to be moved:				
◆ does it present a suitable area to grip?				
◆ is it stable?				
◆ does it obstruct vision?				
◆ is it difficult to stop?				
◆ unsuitable for the characteristics of the load?				
◆ unsuitable for the floor covering?				

(Continued)

TABLE 13.7 (*Continued*)
Checklist to Assess Pushing and Pulling Operations

◆ difficult to steer?				
◆ damaged or defective?				
◆ are brakes effective if present?				
◆ inspected and maintained regularly?				
Consider the **working environment** – are there:				
◆ constraints on the adoption of suitable postures?				
◆ confined spaces/narrow passageways?				
◆ damaged or slippery floors?				
◆ ramps/slopes/uneven surfaces?				
◆ tripping hazards?				
◆ suitable lighting conditions?				
◆ suitable thermal conditions?				
◆ strong air movements?				
Consider **individual capability** – does the job:				
◆ require unusual strength or capability?				
◆ cause a problem for anyone with a health problem, physical impairment or learning difficulty?				
◆ cause a problem for pregnant women?				
◆ require special information/training?				

(*Continued*)

TABLE 13.7 (*Continued*)
Checklist to Assess Pushing and Pulling Operations

	YES	NO	
Equipment:			
♦ Is clothing or personal protective equipment a hindrance?			
♦ Is additional personal protective equipment required?			

TABLE 13.8

A 'Tick-List' of General Elements to Be Considered During Any Assessment

ELEMENTS TO BE CONSIDERED DURING ASSESSMENT	
WORKSTATION DESIGN: Desk features: • Shape • Height • Surface area • Cable management Chair features: • Height adjustment • Backrest height and tilt • Seat length, tilt, area • Five-star base • Armrests Sitting/standing work Footrest Accessories	**WORK EQUIPMENT:** Desktop screen(s): • Height/Distance • Priority usage Keyboard: • Distance • Frequency of use Input device: • Type • Distance • Frequency of use Mobile devices: • Type • Manner and location of use • Frequency of use Trolleys
WORK ORGANISATION: Work rate Duration of activity Rest-breaks: • Timing • Length • Flexibility Self-autonomy Reward system Overtime Training/supervision	**TASK DESIGN:** Nature of task: • Varied • Repetitive • Prolonged • Manual handling Force Self or system paced Postures Static muscle work Dynamic muscle work

(Continued)

TABLE 13.8 (*Continued*)

A 'Tick-List' of General Elements to be Considered During Any Assessment

ELEMENTS TO BE CONSIDERED DURING ASSESSMENT	
WORKING ENVIRONMENT:	**THE WORKER:**
Temperature: • Air temperature • Surrounding surfaces • Air movement • Humidity • Air quality	Bad habits Training level Skill base Age
Lighting: • Level • Location	Gender Health
Noise: • Source	Experience/maturity
Vibration: • Source	Disability Psychological issues
Floors: • Condition • Levels	**OTHER:** Quality control
Space: • Over-head clearance • Surrounding area • Foot and knee clearance	Process failure Seasonal demands

References

Chiasson, M. E., Imbeau, D., Major, J., Aubry, K., and Delisle, A. 2015. Influence of musculoskeletal pain on workers' ergonomic risk-factor assessments. *Applied Ergonomics* 49, 1–7.

David, G. C. 2005. Ergonomic methods for assessing exposure to risk factors for work-related musculoskeletal disorders. *Occupational Medicine* 55(3), 190–199.

Laestadius, J. G., Ye, J., Cai, X., Ross, S., Dimberg, L., and Klekner, M. 2009. The proactive approach–is it worthwhile? a prospective controlled ergonomic intervention study in office workers. *Journal of Occupational and Environmental Medicine* 51(10), 1116–1124.

Pillastrini, P., Mugnai, R., Bertozzi, L., Costi, S., Curti, S., Guccione, A., Mattioli, S., and Violante, F. S. 2010. Effectiveness of an ergonomic intervention on work-related posture and low back pain in video display terminal operators: A 3 year cross-over trial. *Applied Ergonomics* 41(3), 436–443.

Pillastrini, P., Mugnai, R., Farneti, C., Bertozzi, L., Bonfiglioli, R., Curti, S., Mattioli, S., and Violante, F. S. 2007. Evaluation of two preventive interventions for reducing musculoskeletal complaints in operators of video display terminals. *Physical Therapy* 87(5), 536–544.

Robertson, M., Amick III, B. C., DeRango, K., Rooney, T., Bazzani, L., Harrist, R., and Moore, A. 2009. The effects of an office ergonomics training and chair intervention on worker knowledge, behavior and musculoskeletal risk. *Applied Ergonomics* 40(1), 124–135.

14

Case Studies

14.1 Introduction

This chapter contains a series of case studies based on actual projects carried out at a variety of commercial premises involving computer operators. The intention of including this chapter is to provide insight into the typical problems encountered in many organisations and the options available to tackle them. In most cases, simple low-cost solutions were all that was required.

14.2 Case Study A

A purpose-built call centre had a dedicated sales team who interacted with current and potential customers through an online chat facility accessed on the company's website. The advisors would sit in front of their desktop computers and wait for the system to alert them to the fact that a customer had requested a chat and the customer had been assigned to them. The alert took the form of a flashed message at the bottom of the screen and a buzz being emitted.

During the chat, the customer would usually ask the advisor about the costs of the company's services and about comparisons with the service they were already using with another company. A conversation would be carried out between the advisor and customer for a period of about 15 minutes, on average. Some customers might require one minute to get the information they needed, other customers might chat for up to an hour, if they were slow at typing or if they left their computer to collect other information needed to continue the discussions. The advisor could be assigned three separate customers simultaneously.

During the course of the chat, the advisor might have to ask a series of questions so that they could give the customer the information they needed.

Providing the relevant information usually required the advisor to open a series of screens on the company system so that they could provide specific information to the customer.

Throughout the chat, the user inputted text using the keyboard, inputted numbers using the number pad embedded in the keyboard and used the mouse to open and close screens and to copy, paste, and scroll through information.

One of the sales team started to complain about pains in their wrists and forearms. Within two weeks, the sales advisor was signed off work with an upper limb disorder, and they did not return to work for a month. The individual in question had been performing the same task for six years without any apparent difficulty before.

An assessment of the advisor's workstation showed that they had a fully adjustable chair in good working order. The chair had the potential to allow the advisor to sit at an appropriate height relative to the desk and keyboard. A footrest was provided for use. The desk was set at 720 mm in height and was 800 mm in depth, so was considered acceptable for the work that was performed. The screen was located on an articulating arm that allowed the screen to be adjusted for height, tilt, and distance. The keyboard and mouse were standard, wired equipment.

At intervals throughout their employment, the advisor had completed a number of online self-assessments. Prior to being allowed to move to the actual assessment questions, the advisor had to work through an e-learning package which measured understanding at certain points throughout the tutorial. The training was detailed and would have conveyed an accurate and thorough understanding of how to adjust workstation furniture and equipment and how to achieve a comfortable working posture.

Overall, the workstation furniture and equipment would have been considered acceptable for the user and the work that was performed at the workstation. However, it was decided that on the advisor's return to work, they should be provided with alternative equipment that might assist them to carry out their work and remain comfortable. To that end, they were provided with a smaller keyboard with a pull-out number pad which could be stowed within the body of the keyboard when not in use. The use of this style of keyboard meant that any input device to the side of the keyboard was not located as far away from the advisor than was necessary. The user was given a trackpad mouse to use instead of a standard mouse. This eliminated the need for the user to grip anything for extended periods throughout the length of their shift. This also eliminated the need to adjust the natural, or neutral, resting posture of the hand to one that could hold the mouse.

Analysis of customer numbers visiting the website showed that the requests for chats increased markedly when a campaign was being run to advertise new products and services or when special offers were made available. However, contingency plans were not put in place to increase the manning levels in the sales team. Prior to the advisor complaining of upper limb pain, the workload for advisors had increased 300% following a new campaign.

A full review was made of the working practices in force within the office as a whole and within the advisor's specific team. Generally, every user within the office was permitted to take one 30 minute lunch break and two 15 minute rest-breaks throughout their shift. In addition, they were permitted to leave their desk at intervals to visit the restrooms or to get a drink. As a policy, that was considered ideal.

It became evident during the review that the manager of the particular team in question was influencing their behaviour, including the way in which they worked and how they took breaks. Being a sales team, the emphasis was on achieving as many sales as possible. Every morning, the manager emailed the team with a breakdown of the performance levels of each individual within the team on the previous day. It was evident from the table that was emailed who had completed the highest number of chats and who had completed the highest number of sales. The manager praised the best performers and specifically mentioned team members who were considered to be underperforming. Although there was no specific target to meet, the team had a clear understanding of what was expected of them. It also transpired that the sales team earned commission on each successful sale, which operated as an incentive for them to work as fast as they could and to deal with as many chats as they could during a shift.

A further review of how advisor work behaviours were being influenced revealed that the manager also provided a daily breakdown of how much time each member of the team spent away from their desk, or how much time they spent away from chatting with customers even if they were still at their desk. Those who remained at their desk for the longest periods of time and those who spent the most time chatting to customers were singled out for praise, whereas those who regularly left their desk were highlighted by the manager as having to improve their chat-time, i.e. the time spent chatting to customers.

It was evident that the feedback provided by the manager on a daily basis was discouraging advisors from leaving their desks regularly and discouraged them from taking 'micro-breaks' when at their desks.

It was concluded that although the sales environment was one where the focus was on selling services to customers, there had to be a ceiling on what an advisor was expected to achieve so that they did not constantly strive to work longer and harder. The ceiling was not the target; it was simply a cut-off point so that the advisor recognised that they needed to work within a zone that would not result in them developing ill-health. The software in use was updated so that a warning was flashed at the user if they had spent too long chatting with customers without some form of break. Break taking was subsequently actively encouraged.

14.3 Case Study B

The control room of a utilities company had recently been redesigned to accommodate engineers and controllers who had to monitor the flow of energy within and between different areas of the country. The redesign also included the provision of new software that enabled the engineers to complete their work. It could be used for up to 7 hours by the day and evening shifts, 10 hours by the night shift, and for 12 hours during the weekend shifts. It was understood that, potentially, the new software system could be used for extensive periods of time, sometimes with little in the way of interruption, and given the nature of the business, it was essential that it was designed to allow the engineers to interact with the system easily, quickly, and accurately.

Before a full switch-over between the old software and the new software had taken place, the engineers reported that they were making significant mistakes that could have serious implications for the company, in addition to creating catastrophic environmental harm.

To assist in identifying why the engineers were having difficulties in interpreting the information that was being presented to them on the screen and why they were making errors in their on-screen selections, a comparison was carried out between the old software system and the new.

In terms of layout, the data presented on the old system was in a well-defined block, which did not fill the screen fully and did not, therefore, become visually overwhelming. The data was presented in columns which were clearly distinguished from each other by uniform space that reflected the number of digits contained in each column and, as a consequence, there did not appear to be 'wasted' space to the left and right of the digits within the columns. Headers at the top of each column were concise and did not exceed the maximum number of numerical characters in the columns below. Each line of data was separated visually by a distinct and uniform space above and below the digits. Command boxes were separated from the main columns of data, and they are positioned at regular intervals across the screen and were in capitals which made them distinct from each other and eliminated any visual confusion. It was concluded that this layout would have facilitated the decision-making process.

The old screen display employed a 'traditional' layout in terms of colour combinations for the main body of data; yellow was used against a black background which increased legibility.

The size of the characters also enhanced the legibility of the text. The font in use employed a number of serifs; 'serifs' being the small finishing strokes on the end of a character, for example, as shown in the following Times Roman letter i.

Terms or titles used at the beginning of lines were brief and the use of capitals and lowercase was standardised along and between lines, and it matched mixed-case expectations.

There were relatively few pages to navigate through, and the pages were numbered, which the users felt speeded up the process of finding the page they needed.

As a result of the combination of these factors, the old system displayed information in a format that made it easy to read at a normal seated viewing distance, and it ensured that it was read rapidly and accurately. This was considered essential as the old system relied on manual intervention by the controller or engineer.

A subsequent review of the new system identified that there were many features that were likely to interfere with the ease with which the data could be viewed on the desktop screen. The displayed pages were visually 'noisy', and the tables of data were overwhelming. The extent of the information presented was such that significant effort was required to move around each page and to identify the data of importance.

There were a number of factors that, acting in combination, made the data much more difficult to read. This would have resulted in users extending the length of time they attended to the information on the screen as they deciphered the text or content and, as a consequence, it took longer for them to make a decision and act on that decision.

The display was such that it was also more likely that users would lean forward in their seats and squint at the text in an effort to increase its legibility.

For regular text, the starting font size was 11. The starting display title font size was 22, and this was also in bold. Although a font size of 11 may have been adequate in some applications, the perceived size was dependent on the actual typeface used. The text was perceived to be small and illegible by a number of users. It was possible to use the zoom facility to increase the presentation size of the text, however, to do so resulted in the table under scrutiny becoming so large that it moved off the screen, and additional scrolling was subsequently required both up and down and left and right so that the controller could view all of the relevant data. This resulted in additional mouse use for the controllers. The aim should have been to minimise the extent to which the mouse was used.

Because the new system employed sans-serif fonts, the letters did not have small finishing strokes. The individual letters, therefore, tended to have less line width variation than serif fonts. Software designers often use sans-serif fonts because they appear to suggest simplicity and minimalism and are commonly used for display of text on computer screens. However, the context in which sans serif characters are employed, i.e. font size and colour combination, is significant as it has an impact on their readability and legibility.

The system employed green for both characters and line differentiation, i.e. alternate banding in data tables. In some instances, the system used green on green, which did not provide a sharp contrast. Blue was also used for text, as was black. User trees tended, in the main, to be in black, and they tended to be more legible than the actual pages that they led to upon selection. The tables, which were the dominant feature of the system, appeared to be presented in the main with green and white banding. The overall impression of the screen was that some colours appeared washed out even in the relatively low-level lighting in the control room. Other colours were introduced to some critical screens in an attempt to make the information more readable.

The menu tree of the new system consisted of 1000 pages, which should not have been considered an accessible system that allowed for quick and easy navigation. In a bid to speed up the use of the system, a user tree for specific roles was created so that there were fewer pages to navigate through to get to the desired location within the system. A page numbering system was created which resulted in more rapid navigation.

Navigating through the pages of the new system did not always follow a logical route, and there was a lot of repetition, which resulted in the user getting lost or opening up the wrong page. This situation was compounded by the fact that a number of pages had similar headings, which could result in the user opening the wrong page. Given the number of pages available for navigation, this increased the potential for significant mouse use.

There were certain screens which were worse than others in the new system, such as one table that had green and white banding and text in pink, dark green, red, and blue. The overall effect was one of lack of clarity. Users also had significant difficulty due to the fact that they consistently had to scroll left and right to be able to visually capture all of the information required across one section of the table. This was due to the fact that the data was spread over an extended area, and it was presented in columns with a width that was not commensurate with the number of digits presented in the column. In other words, there appeared to be a lot of wasted space in the tables, and much of this was driven by the size of the headings of each column.

Other examples of wasted space included tables illustrating time periods where the seconds were recorded as ':00'. As there was no requirement for seconds to be included in the table, this detail should not have been incorporated into it, as, by doing so, it increased the space requirement for the column. In other pages, space was allocated to the addition of 'GMT' in columns relating to time which was unnecessary. A similar issue occurred when numbers ending in '.00' were entered into the table. The addition of '.00' unnecessarily increased the space requirement of each column, which resulted in the table extending beyond one easily read screen.

A significant amount of identical information was included in single tables, and the user had to hunt for the difference, which was not highlighted in any way. This did not enhance the ease and speed with which specific, critical information could be found.

When comparing columns of data, users found that the information they needed was not always in columns that were either adjacent to each other or even in close proximity to each other. The system was altered so that, at least in the short term, users could move the columns so that they were side-by-side, which made the scanning of information easier. The difficulty that could not be overridden once this was done, was that when the system refreshed, the columns would go back to their original position. The system was designed to refresh frequently so that it could update the data. The system therefore limited attempts at making it more user-friendly.

There were minor incongruities in the system, such as there were lines of information presented in tables where some lines finished with a full stop and others did not. This highlighted a lack of consistency across the system. This could, potentially, raise questions in the user's mind over whether all the data has been displayed or whether some was missing.

There was a lack of consistency in character use within single lines of text, with a mixture of words in capitals and words in lowercase being used. This made some text (i.e. the capitals) move forward in the display and others (i.e. the lower case) recede to the background, which interfered with the perceived unity of the sentence.

In addition to the difficulties associated with the display characteristics, there also appeared to be an issue with what the system could do in terms of meeting the user's need for information. It did not present the information in accurate detailed graph form, as was the case with the old system. Graphs on the old system provided an immediate overview of events within specified parameters and did not require interpretation or translation, which is what occurred when tables of numbers were analysed in the new system.

Finally, a number of experienced controllers commented that the new system did 'odd things' which they did not understand, and they did not know why it was doing it. This was a concern given that the controllers worked in a safety critical environment.

At the end of the review of both the old and new systems, it was concluded that the software designers had created the screen displays on the assumption that the data would not be interrogated frequently or for long periods of time. However, that did not reflect the nature of the work carried out in the control room. It was clear that the interaction between the controllers and the data being displayed on the screens was extensive and time consuming. It was also apparent that controllers were relying heavily on 'tricks' to be able to perform their normal function within the control room.

As a means to rectify the situation, and so that appropriate 'refinements' could be made to the new system, the software designers were invited to spend some time with the controllers. They had not visited the control room nor spoken to any controllers previously. The aim of inviting them to the control room was for them to observe how the controllers interacted with the system and for the designers to understand what the controllers were trying to achieve. What they had originally designed was quite cumbersome to use and interaction with the system was time consuming. There were dual concerns over legibility and usability, and the use of colour and choice of character styling had resulted in text that was difficult to read. The system was unwieldy and difficult to navigate through, and it did not perform all of the activities that the users required to enable them to complete their work.

The software designers altered the new system so that it was more user-friendly. They took time to understand basic display design principles so that more suitable colour combinations were employed. They incorporated basic ergonomics/human factors principles that enabled the characters, and, thereby, the data/text, to be read quickly, easily, and accurately.

The software designers also drew upon current applicable standards, of which they were previously unaware, and this provided guidance on appropriate design characteristics.

The users were provided with training in the use of the updated system, and they were introduced to the means by which the system could be manipulated easily so that the data was presented in a more usable format for them at any specific moment in time. As the users became more familiar with the updated system, they became faster at navigating through sections of relevant data and through fewer pages.

14.4 Case Study C

An officer worker reported experiencing discomfort in his right shoulder and both upper arms, and he indicated that his lower back and neck were painful sometimes. He also had some stiffness in his hands. He had been experiencing the discomfort for about one year, and it had not resolved following a complete change in workstation furniture six months previously.

As a means to understand exactly what was wrong with the user, he was sent to see a consultant who did an MRI scan, and they identified a muscle problem in his right shoulder. Once he had physiotherapy for that specific problem, it resolved. Following ongoing problems with his back, the user was sent to a rheumatologist who did a full back scan, and he was diagnosed with a long-standing scoliosis. The scan also identified a slight bulge in a disc in the neck. Physiotherapy was used to manage this problem.

The user completed a series of body discomfort questionnaires, and it was found that he became more uncomfortable as the day and week progressed. He was usually comfortable first thing in the morning, but worse by the end of the day, and significantly worse by Thursday. It transpired, from analysis of the daily record of work he was asked to keep, that, on some days, he had no meetings, which meant that he could sit for long periods of time at his workstation without getting up. The need for him to leave his desk at least hourly, even if he had no specific reason to do so, was emphasised, and the user was persuaded to set a reminder on his phone, as it was evident that he got absorbed in his work and simply forgot about getting up.

The body discomfort questionnaires also showed a spike in neck pain after he attended meetings. Investigation into activities considered typical of meetings identified that the user worked with a tablet throughout the meeting to take notes. This was supported on the desk surface and caused him to work with his head and neck tilted forwards. He was advised to use a laptop in place of the tablet, if he had to take notes at all, as this would allow him to adopt a better head position, albeit not as good as he could adopt with his desktop screen.

An assessment of the workstation set-up showed that the user, who was 5′ 10″, was sitting very low relative to his desk and keyboard. This resulted, in particular, in his right arm being raised to use the mouse. This was rectified so that he sat at a more suitable height for his stature. In addition to raising the chair, the seat on the chair was lengthened (i.e. slid forward) so that it offered better support. The backrest was repositioned to give improved support to the lower back, and the armrests were lowered so that the chair could be brought closer to the desk edge and so that the user's shoulders were not artificially raised by resting on armrests that were too high. The user was given a small keyboard which did not include a number pad, as it was not used on a daily basis for his role. This ensured that the mouse could be brought closer to his sitting position. A separate number pad was made available for the occasions when he had numbers to input.

It was evident that the user had moved his chair, pedestal drawers and equipment towards the right of his desk, and he reported that he had done so to avoid a draught which he felt on the left side of his desk. This resulted in him turning his head to the side to view the left of two screens, which were mounted on a supporting arm that had not been moved to the right of the desk along with everything else. The head was rotated by about 30°–40° left of straight ahead. The angle of his head was also influenced by the fact that the left screen had been rotated inwards at a greater angle than the right in an attempt by him to make it easier to read. The draught was dealt with by adjusting the angle of the vents, and all of the equipment was moved back into a better position.

To prevent any additional aggravation of his back, the user was advised about being careful when manually handling objects, particularly when carrying his laptop to and from work. To that end, he agreed to use a rucksack, rather than a bag slung from his right shoulder, which he wore over both shoulders and fastened all of the straps to stop it from moving as he walked.

As a result of the ongoing discomfort caused by his scoliosis, the user was provided with a sit/stand desk, which was electrically powered. He was able to stand and sit down when he felt the need, and he made these changes several times an hour. He completed a series of body discomfort questionnaires at intervals throughout his employment so that his levels of discomfort could be monitored.

An overview was made of the user's activities outside of work, and it became clear that he played golf for about eight months of the year and which he did about twice a week. His symptoms appeared to be at their worst shortly after he resumed golf after the four-month winter break. The user was advised about the care needed when resuming physically demanding activities after a long break. He was made aware of the need to build himself up gradually so that he could become fit enough to play without aggravating any underlying condition. He was also advised to have his swing analysed by a suitable professional, and he subsequently had coaching for his posture and swing, which resulted in a more fluid swing and less discomfort afterwards.

14.5 Case Study D

A closed circuit TV (CCTV) room located in a purpose-built office block was used to monitor activities in the town centre. The purpose was to monitor shoppers and to identify and track any individuals considered to be acting suspiciously or in a threatening manner. The CCTV room was manned by two operators at all times, and three sets of operators worked successive eight-hour shifts so that there was 24-hour coverage in the room. It was recognised that the operators' work was quite stressful. For instance, they might have to track a mugger across numerous streets using the various cameras situated around the town, having witnessed an attack on the monitors. They would have to relay the information back to the police so that the mugger could be apprehended before they disappeared from view. Operators were complaining about one-sided neck and shoulder discomfort, particularly if they used the same workstation on every shift.

The single-span workstation used by the two operators (illustrated in Figure 14.1) was constructed so that it had two symmetrical halves, with each half being used by one operator. The work surface was shaped so that it

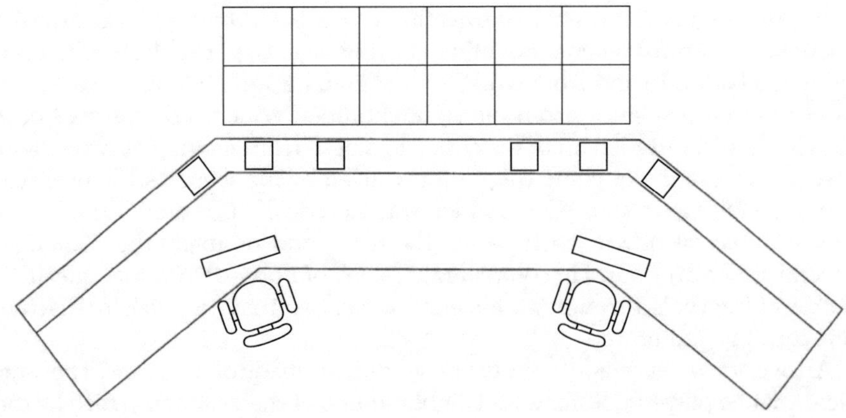

FIGURE 14.1
Illustration of CCTV workstation.

provided each operator with their own L-shaped section. The point at which there was a bend in the surface so that it formed the L-shape was not a smooth, rounded curve, but instead was angled at about 120°. The operators tended to locate their keyboards at this point, and the leading edges of the desk, which formed the angle, tended to push them back away from the desk. This caused them to work at a greater distance from their keyboards.

The work surface was about 400 mm in depth. The control console, which contained six screens, was located at the rear of the work surface that formed the middle. The depth of the work surface resulted in the screens being too close for comfortable viewing. As they were fixed within the confines of a solid console, they could not be pushed back. The location of the screens resulted in the operators moving their chairs back away from the leading edge of the desk so that they sat at a more comfortable viewing distance; however, this simply increased their reaching distance to the keyboards. The location of the console also affected the clearance under the desk, because it was an upright unit that was floor-mounted, thereby providing clearance from front to back under the surface of only 400 mm. This was insufficient to allow the operators freedom to move and change position easily.

The six screens in the console to the rear of the work surface were positioned so that the most frequently used screen was in the outermost position in the console—that is, to the left of the bend in the desk for the operator on the left, and to the right of the bend for the operator on the right. This screen was watched as the operators monitored an incident or situation in the town centre. The screen positioned on the inner side of the bend in the desk was used less frequently as a second reference. In other words, as they watched someone on screen one, they might be able to predict that they were about to head down a particular street and out of view of the camera displaying them on screen one. While monitoring them on screen one, the operator might prepare a camera on the street the subject was likely to move onto and display this street on the second screen in advance of them moving onto it. The third screen was used infrequently, mainly to freeze-frame a picture of an individual so that the operator could provide a more detailed description of their appearance to the police.

Although the two screens used most frequently were positioned directly in front of each operator, one of them was used significantly more than the other. As a result, the operators had to look to either the left or right of their seated position, depending on the side of the workstation they were using, to view the screen on the outside of their workstations. This caused them to work with their heads turned to one side for extended periods.

The work surface was 65 mm thick. This limited the extent to which the operators could raise their chairs before their thighs came in contact with the undersurface of the desk. The added effect of this limitation was that smaller operators could not sit high enough relative to their keyboards, and they were sitting too low to see over the console at the rear of the work surface. This was an absolute requirement of their work because a further bank of screens was located on the wall at a distance behind the desk console. They needed to be able to look at every screen mounted on the wall and be able to see them easily.

The bank of screens mounted on the wall behind the desk was set up so that the screens filled the entire wall area behind the double workstation. As a consequence, when the operators scanned all the screens, they did not move their heads equal distances to the left and right. The operator seated to the left of the double workstation looked further to the right, and the operator to the right of the workstation looked further to the left.

The keyboards in use by the operators were not standard QWERTY keyboards. They contained controls used to direct and focus the cameras. These were located in the centre of the keyboard, and the operators tended to lean on the desk in front of the keyboard and extend their fingers to reach the controls. This overreaching of the fingers was likely to contribute to discomfort at the very least. The control was a small lever with a small knob on the top, and this was so small it could only be gripped by the tips of the thumb, fore, and middle fingers. A number of priority usage control keys were positioned to the left of the keyboard, which resulted in right-handed operators reaching across the body midline to operate them. In situations where the operator telephoned the police while monitoring a situation, they clamped the handset between their chin and shoulder while they continued to operate the controls on the keyboard as they tracked an individual with the system.

The chairs in use were very large, with high backrests and head supports. They were fitted with non-adjustable armrests that were bulky and interfered with the positioning of the chair relative to the front of the desk. The lumbar support of the fixed-height backrest did not offer suitable support to the backs of most operators.

After a full analysis of the work demands and the workstations already in use, it was decided to replace the existing workstations. The work surface remained a conjoined unit with two symmetrical workstations. Figure 14.2 shows the orientation of the new workstation. The new orientation allowed the operators to move their heads to the left and right by equal distances when viewing the bank of screens. The leading edge of the workstation was curved in a horseshoe shape in the area where the operator sat. The depth of the work surface was increased to 500 mm. Leg clearance under the work surface was increased to 750 mm. The desk surface was reduced in thickness to 25 mm. Frequently used items were located within 550 mm of an operator's seated position, and this dimension was also used to position one operator relative to another. The aim was to avoid the need to reach more than 550 mm when passing an item, such as notebook, to a colleague. New, fully-adjustable chairs were provided, and operators were given a briefing on the postures to adopt when working. The priority screen was positioned directly in front of the operator in the console, and the second and third screens were positioned at either side of the priority screen. A new keyboard was designed that incorporated a joystick in the centre to replace the lever, and priority keys were positioned more centrally. Telephone headsets replaced the handsets.

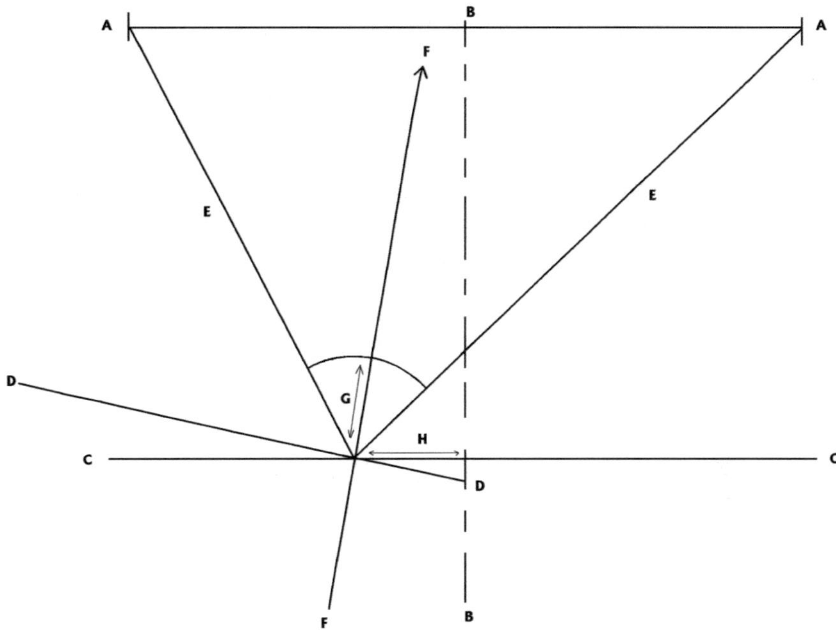

A: Current screen bank.
B: Centre line dividing screen bank and double workstation.
C: Current workstation orientation.
D: Proposed workstation orientation.
E: Outer boundary of proposed viewing angle.
F: Horizontal line of sight for new workstation orientation.
G: 500 mm width of proposed worksurface.
H: 550 mm proposed reaching distance to second operator.

FIGURE 14.2
Illustration of suggested workstation layout.

14.6 Case Study E

The user in this case study was a Communication Operator (CO) who worked for a police force and who handled emergency calls. After she had been working for the police force for about 10 years, she started to experience mild symptoms of pain and stiffness in her right shoulder area which worsened over the next few months. She subsequently attended her GP and was absent from work for six weeks, during which time she was referred for physiotherapy. On her return to work, her symptoms returned and she was signed off work again, this time for three months.

An investigation into what might have prompted the sudden appearance of the CO's symptoms focussed both on her working conditions at the outset of her employment and then just before the onset of her symptoms.

Initially, the CO worked part-time hours, and she worked two early shifts, had two shifts off, worked two night shifts, then had four nights off. She sat at a desk with an adjustable chair and used a touchscreen and computer screen, which were side-by-side in front of her, a keyboard, a wired mouse, and a telephone.

Six months before the onset of her symptoms, the CO moved to full-time work with eight-hour shifts of three early shifts, followed by four late shifts with three days off. She started to work with four screens, a keyboard, a mouse, a headset, a telephone, and a foot pedal, all of which were located on a standard rectangular desk. The cable length on the screens limited their movement so they could not be moved forward nor from side to side. A CCTV screen, which was in addition to the other four screens, and which she did not use in her role, was located on her desk, making it more cramped.

The screen, which was positioned on the left of the CO, was a touchscreen which incorporated the radio system. Observation of the CO when working highlighted that as she was right-handed, she used her right hand on the touchscreen to her left. She reported that she was not dexterous enough to use her left hand on the touchscreen. The posture resulting from using the right hand on the screen located on the left was awkward, as the arm was extended and moved across the centreline of the body. Although the foot pedal located under the desk could be used in the alternative to interact with the radio system incorporated into the touchscreen, so that the CO could talk to a Police Officer, she preferred to use the touchscreen facility instead.

The screen to the right of the touchscreen, and directly in front of the CO when in use, was referred to as the priority screen, and it incorporated the command and control log. The screen to the right of this had the location system, and it was reported that this was referred to rather than interacted with during the work. The fourth screen to the right was also referred to, and the CO did not have to interact with it to the same extent as she did the touchscreen and the screen in front of her.

The touchscreen was swapped with the location system screen so that it was in front of the CO. As use of the foot pedal could, potentially, reduce the number of times the screen had to be touched with the hand, the CO was encouraged to use it more often.

The four screens were placed on separate articulating arms which provided flexibility over how they were positioned. The screen arms also allowed the screens to be tilted and, as they were not fixed to the surface of the desk, they could be moved backwards and forwards. The cables on all of the screens were lengthened so their positioning was not limited. The CCTV screen was removed.

The desk was replaced so that it was adjustable for height and was driven by a motor so the height could be changed easily. If required, the height of the desk could be changed so that it could be operated from the standing position. The desk curved on the left and right sides, which enabled the screens to form a curve on the desk surface.

When the CO made a phone call, she typically clamped the handset between her chin and shoulder, as she had to use her hands on the keyboard and mouse whilst talking. She was provided with a telephone headset in the alternative.

A wireless vertical mouse was provided to allow a more comfortable hand position to be adopted and so that the CO had more freedom over where to position the mouse, as this was used to operate all of the screens apart from the touchscreen.

The CO was given additional training so that she could perform other work within the emergency call handling room. As the role of dispatcher was identified as requiring less use of the touchscreen than the CO role, she was rotated with this task at intervals throughout her shift. She was also able to take brief breaks on an hourly basis so she could leave the workstation.

The shift pattern was changed so that there was a day off between the three days shifts and the four night shifts. This was followed by two further days off before the CO commenced day shifts again.

14.7 Case Study F

A large office block had its own restaurant which had a fully serviced kitchen employing eight members of staff. The business became aware that a number of employees working in the kitchen were taking time off as a result of back pain, which was leaving the kitchen short staffed. This was having an impact on the quantity and quality of food being produced for the office staff.

A review of the work carried out in the kitchen showed that all of the employees in that area were female, were aged 45–63, and all of them performed significant amounts of manual handling. The manual handling being performed was determined by the area of the kitchen in which it occurred.

At the washing station, it was common practice for the large saucepans and casserole dishes to be stored on the floor around the washing up area, and they were left there by the cooks. There was insufficient space on the work surfaces for the large saucepans and only the small saucepans were placed there. As a consequence, every time the larger, heavier saucepans were picked up, they were lifted from floor level. When empty, the larger saucepans could weigh in the region of 6 kg. More often than not, the saucepans and dishes were partially filled with leftover food or water that had been placed inside them to loosen the food residue left after cooking. This increased the weight of the loads being handled and sampling of the weights indicated that the saucepans could increase in weight up to 8–10 kg with the addition of food or water. Clearly, when they were full of food and were moved by the cooks, they were significantly heavier.

Once the saucepans and casserole dishes had been lifted off the floor, they were placed in a deep, commercial kitchen sink so they could be washed. The depth of the sink caused the staff member to lean forward as they washed up, which would have increased the likelihood of the individual experiencing lower back pain.

A dishwasher was used for washing plates, cups, bowls, etc. This was loaded every 5–10 minutes during busy periods. Due to the limited space in the area where the dishwasher was fitted, the dishwasher door could not be left open as it was being loaded. Therefore, the kitchen staff had developed a system whereby the racks were taken out of the dishwasher and placed on the top of it. They filled the racks in this position, then lifted the racks plus crockery into the dishwasher. This was an awkward task due to the precise alignment that was required to re-fit the rack, the bulk of the rack, the weight of the contents, and the fact that the door prevented the handler from standing close to the point where the rack was loaded. In terms of the latter issue, the cooks had a similar problem when removing heavy foods, such as joints and casseroles from the oven, due to the door opening from top to bottom.

Different members of staff cooked different types of food. Some worked on cakes and pastries, which resulted in them handling packs of flour. Although each pack of flour weighed only 1.5 kg, they were delivered to the kitchen area in bulk with 10 bags of flour in each pack. The same occurred with packs of sugar, dried fruit, etc. The delivery driver placed the order on the floor inside the back door to the kitchen and a member of staff manually transferred it to the appropriate section of the store room.

Members of staff handled other heavy items, such as tubs of mayonnaise weighing 12.5 kg, bags of potatoes weighing 25 kg, sacks of onions weighing 15 kg, and drums of oil weighing 18 kg. The oil in the fryers was changed every week, and this resulted in the tray for the oil being lifted out of the fryer and the old oil was poured into a bucket. The drum of fresh oil was carried from a store room and poured into the tray, which was located on the floor, before the tray was lifted back into the fryer. The bucket of old oil was carried to the sink area and was poured back into the empty drum using a funnel, before it was marked as waste and was carried to a storage area.

The store area was small and the shelves were not used in a suitable manner. Packs of drinks were stored on the floor directly in front of the shelves, and if kitchen staff wanted to retrieve a specific product from the shelves, they would either have to move the entire stack of products on the floor to another available space, or they would simply reach over and around it if it could not be moved.

If a specific tray of drink was required from a stack, for instance, sparkling water, the kitchen staff would have to remove all of the other trays of drinks located on top of it before they could pick it up. The trays of drinks removed to gain access to the sparkling water then had to be replaced back onto the stack. A pack of 24 bottles of soft drinks could weigh in excess of

13 kg. The storage arrangement, therefore, led to additional unnecessary manual handling of quite heavy loads.

As a means to reduce the risk from the manual handling, trolleys were provided so that cooks could place dirty saucepans and dishes on them and push them to the sink for washing. This eliminated the need for the cooks to carry the loads and eliminated the need for the person who did the washing up to repeatedly lift from floor level. Cleaned saucepans were returned to the trolleys and pushed back to the cooking area.

The sink was fitted with a false base which, in effect, reduced its depth, and ultimately reduced the extent to which the person doing the washing up had to bend forward when manually washing pots and pans. They were provided with a water jet spray so that less scrubbing of items was needed, which reduced the need to bend into the sink. They were advised to rotate large pots as they cleaned them, rather than reach across the sink to the far side of the pot, which caused them to lean forward.

The dishwasher was relocated so that the door could be opened for periods of time whilst it was being loaded. The racks were left in-situ whilst they were being loaded. The dishwasher was raised off the ground so that the racks did not require the person emptying it to bend down. A drawer was built into the space under the dishwasher, and the tablets and rinse aid and anything else relating to the dishwasher was stored in there so that it did not need to be collected from the storeroom, which was at a distance.

When the cookers were updated, they were replaced with models that had oven doors that opened from left to right, which gave more direct access to the inside of the oven.

Deliveries were taken directly to the storeroom by the delivery driver rather than being left on the floor inside the back door. Bulk packs were opened before the contents were lifted, individually, onto the shelves in the storeroom. A more organised use was made of the shelves, so that unnecessary handling was avoided and heavier items were presented at about waist height. Smaller sizes of products were purchased, rather than catering sizes, so that they were easier to handle. Cleaning materials that were bought in commercial sizes, such as bleach, were purchased in smaller, domestic sizes.

The fryer was replaced with another that could be drained directly into a receptacle to eliminate the need to lift a filled tray out. This also eliminated the risk of oil spillage, which had previously resulted in slipping accidents.

A sack truck was located permanently in the kitchen so that it could be used to move large items or bulk packs of products, rather than the kitchen staff having to move them around manually.

All white goods placed on the floor were retrofitted with castors so they could be moved easily during cleaning.

Bins were emptied more frequently, so that the bags did not become overfilled and heavy.

Staff worked together more often to share the lifting of heavier items.

Index

Note: Page numbers in italic and bold refer to figures and tables respectively.